T0280556

THE INTRINSIC VALUE OF NATURE

VIBS

Volume 59

THE INTRINSIC VALUE OF NATURE

Leena Vilkka

Amsterdam - Atlanta, GA 1997

Cover design by Chris Kok based on a photograph, ©1984 by Robert Ginsberg, of statuary by Gustav Vigeland in the Frogner Park, Oslo, Norway.

∞ The paper on which this book is printed meets the requirements of "ISO 9706:1994, Information and documentation - Paper for documents - Requirements for permanence".

ISBN: 90-420-0325-1

Printed in The Netherlands

To Holmes Rolston III,
with thanks

CONTENTS

EDITORIAL FOREWORD

Leena Vilkka takes us on a critical journey through the forest of distinctions and theories that treat life and nature in terms of values. Such values have been variously proposed as intrinsic, inherent, instrumental, immediate, inherited, absolute, derivative, self-generated, objective, subjective, experiental, personal, biological, social, natural, or global. To this proliferation of ideas, Vilkka brings systematic order and sensitive appraisal. She defines the positions, charts their structures, contrasts their programs, weighs their insights, and corrects their shortcomings.

Throughout the philosophical expedition, Vilkka keeps her eye on the living matter that may evade the nets and snares of value-theory. What matters is life. Life has many biological forms, in addition to that form which has been our central concern as theorists and as animals, namely, humanity. Even the Earth might be thought of as a living organism.

Often the theories of values elaborated by human beings that are applied to life and to nature as a whole give priority to the human species. How could we do otherwise? Even theories that take the extra step of arguing for the rights of other living things may operate by extending a set of human values, as if by generosity, to those who are not human beings. The movement is from human rights to the rights of nature.

Vilkka calls for a change of perspective so that we can appreciate the rights of our species, along with the rights of others, as coming under the broader canopy of the value of life. Thus, the rights of nature may give justification to our human rights.

Vilkka is not asking us to forget our humanity. She helps us to discover a fuller sense of what is truly valuable in life on this threatened planet.

Robert Ginsberg
Executive Editor
Value Inquiry Book Series

PART I

ON BASIC CONCEPTS

One

WHAT IS INTRINSIC VALUE?

1. Introduction

This book is a theoretical inquiry aimed at forming a theory of the intrinsic value of nature. It provides philosophical arguments for animal preservation and nature preservation. The third task of the book is to persuade its readers to realize the importance of its subject-matter. Most philosophers and scientists do not accept the language of intrinsic value in relation to the non-human world. This book questions two common assumptions: (1) only human beings have intrinsic value; (2) human beings are the origin of all values. I am critical of anthropocentrism, although I propose a form of ecohumanism, which recommends an attitude of respect for animals and nature. The ethical goal, which I have espoused, is the well-being of the people on Earth and the well-being of the Earth and its diverse life forms, plants and animals.

Intrinsic value is a notoriously difficult concept. The debate surrounding this key concept has been confused. Among the authorities in the field of intrinsic value, no agreement has been reached about basic terms, concepts, and ideas. Some good books on intrinsic value exist, although the experts have different views and few attempts have been made to reconcile their individual terminologies (like inherent value, inherent worth). The term "intrinsic value" has become broad and imprecise. Virtually anybody or anything can be the location of intrinsic value or the object of intrinsic valuation: individuals, universals, and wholes, or people, animals, plants, and ecosystems. We need to clarify and classify different forms and meanings of intrinsic value.

The search for intrinsic or natural value is the ultimate ground of nature-conservation philosophy.[1] Anthony Weston claims that "it becomes easy to justify respect for other life forms and concern for the natural environment, and indeed many of the standard arguments only become stronger, once the demand to establish intrinsic values is removed."[2] I defend the contrary view: we have a substantial need to prove the assumption that the natural world is of worth in itself.[3] Intrinsic value in nature is the most fascinating problem of nature-conservation philosophy from the theoretical point of view. The position that we should not attempt to establish the intrinsic value of nature cannot be successful from a practical point of view. We do need a notion of intrinsic value to justify environmental policy for treating the non-human world. Several instrumental human interests occur in the environment: economic, ecological, scientific, religious, cultural, aesthetic, and leisure interests. Non-instrumental human interests in nature are also strong. The willingness of nature activists to preserve nature for its own sake should not be ignored in environmental practice and theory. We need the intrinsic-value view of nature to form a complete frame of nature's values.

The basic interests of this book concern animals (other than human beings), living beings, and ecosystems. I offer a naturistic (nature-istic) argument. A humanistic (human-istic) approach would be at stake, if the topics of human beings were made central. My argument is based on the classification between naturocentrism and naturogenic value. Anthropogenic value in nature is a form of value generated by human beings and centered on nature. Naturogenic value is generated by nature. The argument has three components: (1) the ethical and ontological types of intrinsic value in nature (zoocentric and zoogenic value, biocentric and biogenic value, ecocentric and ecogenic value), (2) the specific ideals of intrinsic values (well-being, biodiversity, and beauty), and (3) the generic levels of value in nature (consciousness, livingness, and wholeness). I argue that animals and nature have sufficient intrinsic value that in decision contexts animals and nature should win, since their values have priority over human values.

This book is divided into three parts. Part I develops basic concepts and distinctions of intrinsic value. In Part II, basic nature-centered concerns are discussed in detail. Part III is focused to three themes: the origin of value, the problem of priorities between human and natural value, and rights for animals and nature. The purpose of Chapter One is introductory. It defines the basic distinctions and issues posed by the problems of the intrinsic value of nature. In Chapter Two, I distinguish between intrinsic, instrumental, and systemic goodness in nature. Chapter Three continues the classification of intrinsic value in terms of good. Four basic groups are: 1. intrinsic value in terms of experienced-value, 2. intrinsic value in terms of ends-value and of self-value, 3. the group of subjective intrinsic value (subject-centered intrinsic value), and 4. the group of objective intrinsic value (object-centered intrinsic value). Finally, intrinsic value is understood in terms of respect and love for animals and nature. Activists on behalf of animals and of nature express their willingness to preserve nature for its own sake by using the language of intrinsic value.

Part II concentrates on three basic topics of nature: animals, life, and nature as a whole. Chapter Four offers the animal-centered philosophy called zoocentrism. I focus on animal well-being, animal subjectivity, and animals as sentient beings. Tom Regan's concept of inherent value of sentient animals is analyzed in detail. Chapter Five proposes the model of biocentrism, life-centered philosophy. Basic questions are "What is life?", and "What is the intrinsic value of life?" Some old philosophical issues of life are fruitful in the modern discussion of biodiversity. I introduce the old principle of plenitude, and the concepts of will to live (Arthur Schopenhauer, 1788-1860) and *élan vital* (Henri Bergson, 1859-1941). The present concept of inherent worth is a part of Paul W. Taylor's biocentrism and is examined in detail. Taylor compares biocentrism to anthropocentrism. Biocentrism regards respect for living nature in the same way as anthropocentrism regards respect for people. Taylor builds his theory of nature's inherent worth on the concept of the good of each living being. His argument is roughly as follows: if an entity has a good of its own, and if it is better that this good be realized than not be realized, then the entity has inherent worth. For a living entity it will be better that its good be realized than not. Thus, all living organisms have inherent worth. This includes the moral argument that animals and plants should be treated in such a way that their good be realized rather than not. Taylor's individualistic approach is principally challenged by holistic nature thinkers, such as Aldo Leopold (1887-1948) and J. Baird Callicott. Chapter Six considers ecocentrism, the philosophy of ecosystems. Its three sub-forms are divided into anthropogenic holism (J. Baird Callicott), biocentric holism (Lawrence Johnson), and physiocentrism (Klaus Meyer-Abich).

In Chapter Seven, I develop my naturistic theory, in the sense of ethical and ontological extensionism and value pluralism. I argue for the existence of intrinsic value in nature, including the intrinsic value of consciousness, life forms, the land, ecosystems, and the earth. This approach develops naturogenic value subdivided into zoogenic, biogenic, and ecogenic value inhering in non-human nature independent of human valuation. The issue of natural beauty is discussed in this chapter. Chapter Eight discusses anthropocentric intrinsic value in nature. Weak anthropocentrists can argue that things other than human beings have intrinsic value, albeit human worth takes priority over the intrinsic value of nature. Ecohumanism is the promising model of weak anthropocentrism in defending both human and natural value.

Finally, I offer three types of argument for nature's intrinsic value in relation to rights: the conceptual points of values and rights, the ontology of intrinsic value and moral rights, and the practical preservation utilities arising from the relationship. In Chapter Nine, I argue that our view of the rights of animals and nature depends on our general view of the world, whether it is anthropocentric or naturocentric. We should extend the idea of natural rights to the rights of animals and nature.

In recent discussions, no comprehensive book has been written against the intrinsic-value view of nature. The present book defends the pluralism of value and classifies basic forms of value in nature. The notion of intrinsic value related to environmental protection is strikingly new in philosophical discussion. We even cannot argue against the view that such values exist in nature, unless we have suitable theories in the field. My hope is that this book inspires the discussion of intrinsic value.

PART I: ON BASIC CONCEPTS

CHAPTER 1	CHAPTER 2	CHAPTER 3
What is the problem of the intrinsic value of nature? What is nature? The instrumental value view of nature The intrinsic value view of nature	Goodness in nature (1) intrinsic, instru-mental, and systemic goodness (2) biological and axiological goodness What is good for nature?	The classification of intrinsic value (1) experienced-value (2) ends-value (3) self-value (4) inner value (5) intrinsic value in terms of respect and love

PART II: NATURE CONCERNS

CHAPTER 4	CHAPTER 5	CHAPTER 6
Zoocentrism (respect for animals) Animal well-being Animal subjectivity Tom Regan's theory of the inherent value of sentient animals	Biocentrism (respect for life) What is life? Biodiversity issue Paul W. Taylor's theory of the inherent worth of living beings	Ecocentrism (respect for ecosystems) The organic world-view Individualism vs. holism Aldo Leopold, an ethical extensionist

PART III: THREE BASIC ISSUES

CHAPTER 7	CHAPTER 8	CHAPTER 9
Ethical extensionism and the origin of value A theory of naturo-centric (ethical issue) and naturogenic value (ontological issue) Naturogenic value subdivided into zoogenic, biogenic, and ecogenic value	Anthropocentrism and priorities Weak and strong anthropocentrism Ecohumanism Which should have higher priority, human or natural value?	Rights for animals and nature Rights as practical tools for animal and nature preservation Anthropocentric and naturocentric rights

The Structure of the Book

2. What Is Nature?

No science or academic discipline can grasp the totality of nature, or claim that it knows best what nature is. Theoretically, the views of nature are dependent upon our world-view based on materialism or idealism, realism or relativism, subjectivism or objectivism, humanism or naturalism. In practice, people have different experiences of nature from their childhood. The real nature is our childhood nature. We may hold nature as a forest area including exciting natural things. Nature is outside and between cities, roads, fields, and houses. Nature refers to forests and wild animals, while the concept of animals refers to sentient creatures living in human societies.

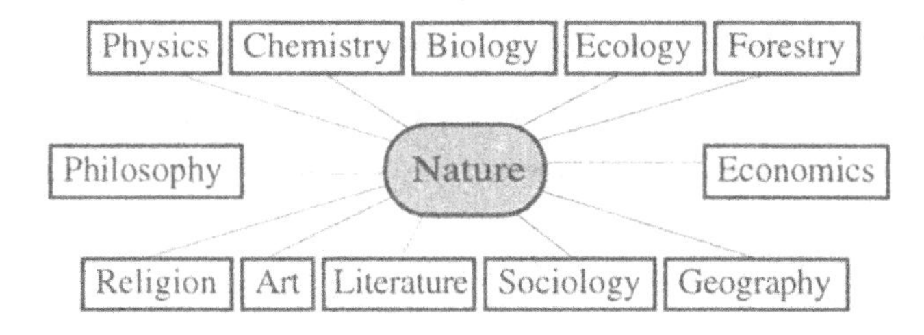

Nature means different things to different people: to the skier, the country-dweller, the hunter. It is a habitat for a variety of animals, plants, and fungi. Nature as the forest has provided the material well-being on which Finland's economic and cultural advancement is founded. Nature is a battle of words between commercial utilization and nature conservation. In the instrumental view of nature, the forest is a source of raw-material for people. Efficient forest management converts natural forest into commercial forest, with limited opportunities for plants and animals to live. In the language of intrinsic valuation, sufficient parts of the forests must be left untouched to provide habitats for animals, plants, and fungi. Nature's values, tangible and intangible, should be preserved. We have no right to exploit natural values when achieving human values. Preserving nature means some harm for people, although the well-being of nature and people is connected in an ideal world. The well-being of nature preserves us. We do not behave reasonably if we harm nature or decrease biodiversity. In the real world, no harmony between human welfare and nature preservation exists.

The topic of animal rights concerns mostly pet animals, meat animals, laboratory animals, and the like. These animals are distinguishable parts of nature. Different criteria and ethical principles are needed for city animals and wild animals. The animal-centered philosophy, zoocentrism, seeks the philosophical criteria and ethical principles for animal preservation. The life-centered philosophy (biocentrism) and ecosystem-centered philosophy (ecocentrism) both promote arguments for nature preservation. These are autonomous but complementary issues for understanding our human relationships with the non-human world.

In this book, I grasp the question of "What is nature?" by subdividing it into the issues of animals, living beings, and ecosystems. What are animals? What is life? What is nature as a whole? The concept of nature refers to animals, living beings, and ecosystems as opposed to human culture and artefacts. Animals are regarded as sentient beings to distinguish them from non-sentient nature. Plants as living beings are the opposite of non-living things. Ecosystems include fauna and flora and the ecological process of nature. My aim is to develop the views of intrinsic value that apply to these three categories. A sentient animal is a cow, a rat, a bird, a snake, a bat, perhaps a butterfly. By non-sentient life-forms we refer to living beings: lower animals, plants, and trees. Ecosystem refers to the biotic and abiotic environment. A forest, a mountain, or a river forms an ecosystem.

Naturocentrism	Focus	Key Concept
Zoocentrism	Animals	Consciousness
Biocentrism	Life	Livingness
Ecocentrism	Ecosystems	Wholeness

3. The Instrumental-Value View of Nature

The tension between intrinsic and instrumental value concerns human attitudes and ethical relations to animals and nature. By the instrumental-value view, I mean the dominating view in the present environmental policy, which takes only the instrumental value of nature (mostly in terms of money) into account. An example is chickens in modern animal farms. They live in small cages, in which their life is controlled by automatic systems for light, feed, and water. These chickens are treated as having their material value as flesh and products for a meat company. The technique and methods of modern factory farms treat lives of animals as instrumentally valuable. The animals have no value in themselves. No respect for these animals, for what they are in themselves, exists. The key question is whether the instrumental-value view of nature is a sufficient ground for responsible human relations to animals and nature. At the shallow level, animal suffering and pollution (when caused by people) may have no connection. By looking at the deeper level, the background of animal suffering and pollution is the instrumentalist view, in which animals and nature are seen as intrinsically valueless raw-material for commercial purposes. The instrumental-value view of nature is challenged by the intrinsic-value view, which argues that effective restrictions are necessary to combat the inhumane and destructive ways that we treat animals and nature in our commercial and industrial age.

The instrumentalists remain a minority among philosophical theoreticians. Bryan Norton has defended the transformative value of nature as an instrumental value.[4] He claims that the instrumental-value view is the sufficient and reasonable ground for nature conservation, nature having several instrumental values for people. I call this the wide instrumentalist view, because it opposes the narrow instrumentalist view according to which all values in nature are in terms of money. In a commercial or industrial context, nature is a pure instrument in terms of resource for the economic purposes of people. Nature is held to be a raw-material. Instrumental values are understood in terms of utility, and intrinsic natural values are ruled out of existence. Human relations to nature are technical, industrial, or commercial. The non-human world has only monetary value.

I challenge the instrumental attitude, both a wide and narrow instrumentalist view, to animals and nature. My challenge for environmental policy is that the varieties of non-instrumental value in nature need to be taken into political account. One of my claims is made in terms of animal rights. Animals as sentient creatures of intrinsic value should be respected, and their rights should be confirmed by laws. We should not escape the question of how the non-human world should be treated. And we should decide to treat it as not only commercially or productively valuable, but as ethically valuable as well. We need to have an attitude of respect for nature. My theory of intrinsic value in terms of respect for animals, plants, and ecosystems opposes the narrow instrumentalist view, which sees all natural things as raw-material and instruments for human purposes.

Instrumental reason has been critically discussed by philosophers of technology, including such well-known authors as Herbert Marcuse[5], Jacques Ellul[6], Lewis Mumford[7], Theodor Adorno and Max Horkheimer,[8] E. F. Schumacher[9], and Carolyn Merchant[10]. The philosophical critique of technology questions the instrumental relation of people to the environment. The current thinking about nature is recognized as based on technological and commercial values. Modern science reveals facts about nature which can be used to develop technologies for manipulating nature. On this view, nature is a thing to be utilized. In Western culture, nature has been a means exploited for human ends. Material success in exploiting nature is a background for an instrumental world-view. The narrow instrumental value of nature is close to "use value"; nature is a resource, raw-material, or tool. Yet nature should be preserved and protected because of its wider instrumental value for human welfare. Some instrumental values, such as recreational and experiental values in nature, are compatible with intrinsic value. We should link together the instrumental and intrinsic-value view instead of regarding them as opposite views. I wish to couple them for a viable environmental policy, although I discuss at length the theoretical problems and concepts in terms of the intrinsic-value view of nature. Another possibility is to argue that the intrinsic-value view is more fundamental than the instrumental-value view. To break away from instrumentality is necessary to build the alternative view, which respects nature as a value in itself.[11] The argument is that instrumental values can be related to things or states which are instruments or means to achieve things or states which are of intrinsic value. Intrinsic value refers to something which is (or has) value in itself, and instrumental value cannot exist without an underlying reference to intrinsic value, for there would be an infinite regress of instrumental values with nothing valued or perceived as value in itself. Hence, intrinsic value must be the most fundamental form of values.[12]

A. The infinite regress of instrumental values (value for something):
Instrumental value —> instrumental value —> instrumental value —> instrumental value ...

B. Instrumental values lead to something, which is of intrinsic value (value in itself):
Instrumental values —> intrinsic value

The instrumental-value view of nature is a crucial argument for nature conservation. On another perspective, the problem of which things have instrumental value for nature belongs to my interests. Two different topics are (1) the instrumental goodness of nature for people, and (2) the instrumental goodness of some natural thing for some other natural thing. The second is a novel issue. Typically, the instrumental-value view is understood as nature being instrumentally valuable for people, while I propose that we should seek which things are instrumentally valuable for animals and nature. We should ask what makes us human beings instrumentally valuable for nature (for more on this, see Chapter Two).

4. The Intrinsic-Value View of Nature

In this section, three standard classifications are elaborated: (1) intrinsic and instrumental valuation, (2) intrinsic and extrinsic value, (3) objective and subjective value. First, in the distinction of intrinsic and instrumental valuation, intrinsic value is an end-value, referring to what is valuable for its own sake, as the opposite of instrumental value. Secondly, in the distinction of intrinsic and extrinsic value, intrinsic value is an inner value of an object in terms of value in itself, as the opposite of extrinsic value. Thirdly, objective intrinsic value is defined as a qualitative property of an object. Finally, the topic of moral standing is subdivided into moral considerability, moral significance, and moral priorities. Ethical extensionism is offered as an answer of which entities should have moral standing.

A. What Is Intrinsic Value?

The term "value" is closely related to an older and classical philosophical term, "good." What is good and what is value cannot be totally distinguished from each other. Some differences between value and good can be found. The term "value" is a noun for which the adjective is "valuable" or "valued," while the term "good" is normally used as an adjective. "Valuable" can mean good to possess or good to use, and thus suggests that a thing is valuable for somebody. "Value" has the verb form "to value" or "to evaluate," while "good" does not have the verb form. Value in terms of what is good can be classified in three basic forms:

(1) Good of a kind
(2) Extrinsically good, which can be subdivided into (a) instrumental good and (b) contributive (or transformative) good.
(3) Intrinsically good.[13]

Good of a kind means that something is good in a some respect. Apples are good when they are good-looking or tasty. Knives are good when they are sharp. Extrinsic goods means values which are derived from something else. Those entities which are means to something else belong to the sub-class of instrumental goods. Those which are necessary parts of a good whole belong to contributive or transformative good. Intrinsically good are good in themselves.

What does it mean to say that value is intrinsic? A Thesaurus lists the following words with meanings close to the word "intrinsic": "inborn, inherent, innate, inner, inside, internal," and "natural." Other words are as follows: "actual, authentic, genuine, real," and "true." Value indicates significance and worth (the case of intrinsic value) and advantage, benefit, and utility in the case of instrumental value.

X has intrinsic value if

(1) it is sought or desired for its own sake;
(2) its value depends on its nature and not its consequences or it has non-derivative value;
(3) an objective, non-natural property inheres in or belongs to it;
(4) it has this value even if it were the only thing existing in the universe.[14]

The first classifies ends-value as opposed to means value. The second refers to the old naturalistic view of values, in which values are understood in terms of the natural (or physical) properties of objects. X having non-derivative value is the view which can be found in the authors G. E. Moore, C. A. Baylis, and C. I. Lewis. The third is classic, Moore's special view of intrinsic value. The fourth refers to some kind of isolated value. I discuss in detail the idea (1) in section C, from the points (2) and (3) in sections D and E. In the following, I discuss briefly the issue (4) by utilizing the problem of the last human being in the world. As a thought experiment, its attempt is to show that non-human and non-sentient beings are of intrinsic value, apart from their existence valued by any sentient beings.

The thought experiment (formulated at first by Richard Routley) runs as follows. Image that all living organisms have been destroyed by nuclear warfare and the last person is the only living being who will die soon. This person has the ability to destroy all diamonds remaining in the world. Will she or he now be destroying anything of intrinsic value? My answer would be that she or he will not. Consider another case where all sentient creatures have been destroyed while plants remain. The last person intends to cut down the last tree, which could reproduce new generations if left to live. Will the last people now be destroying something of intrinsic value? How you respond to the issue depends on whether you have already taken an anthropocentric or naturocentric attitude to the non-human world. For anthropocentric thinkers, a deed is wrong if it does harm to people. For the naturocentric thinkers, cutting the last tree would be a wrong act. An argument for the intrinsic value of the last tree is that trees have their own good which is independent of the human good. This aspect of intrinsic value is ontologically independent of your thinking or feeling (on subjective valuing and objective values see Chapter Seven, on the good of the living beings, see Chapter Five).

The basic conceptual, ontological, epistemological, and ethical questions about the intrinsic value of nature:

1. What does "the intrinsic value of nature" mean?
2. Are there intrinsic values in nature? What are intrinsic values in nature? How do they exist in nature?
3. How do we perceive them in nature?
4. What is their ethical significance to human life?

In a sense "intrinsic" refers to the real world, like "intrinsic facts," such as gravitation, in the language of physics. Gravitation is independent of human mind, although the concept of gravitation is human-dependent. Intrinsic value in philosophy is like truth in science. We may not exactly know what "intrinsic value" is, albeit we can grasp it from different points of view. In the language of truth and intrinsic value, we ask, what are things in their real nature? The fundamental question of the ontology of value is: do we value things because they are valuable in itself or are they valuable because we value them for their own sake?[15] Do we attribute value to nature, or do we discover value already present? This distinguishes between the issue of valuation and value which I consider next.

B. Value and Valuation

The dispute of values is about which sense is primary: value as a noun or a verb. If the verb "to value" is primary, then the noun "value" designates something valuable, something valued, which is the object of a valuing activity on the part of human beings (or other valuing subjects). Value as primarily a noun designates an object in its intrinsic quality, whether or not human beings or sentient beings value it. This dispute appears in anthropogenic and naturogenic theories of intrinsic value. Anthropogenic theories emphasize that value is related to its verb form "to value" or "to evaluate," that is, to human valuation processes. Naturogenic theories emphasize the substantive or adjectival form of value in which intrinsic values are independent of human valuers in nature. The question of valuing and value arises in the discussion of the subjectivity of animals. We need the category of intrinsic value for beings valuing their own lives. I call this the value of self. Intrinsic value is sometimes so called in Finnish. Conscious beings can value different kinds of things and states as intrinsically valuable, and conscious beings can in turn be valued as intrinsically valuable by other beings.

The second distinction should be made between value and things having value. We need both a theory of value and a theory of those things which are the objects of our valuation. The question, "what is value?", should be distinguished from, "what kinds of things have value?" Ethical extensionism primarily seeks answers to the question of what kinds of things have value, by extending the scope of things having intrinsic value from people to non-human nature. If we human beings have the ability to value things, this ability cannot be restricted to the ability to value only human beings. We value many concrete and abstract things. For example, scientists have argued that basic research has its own value (ends-value) regardless of its instrumental utility for human society. And nature preservationists claim that some wilderness areas should be preserved for their own sake, regardless of their instrumental utilities.

The third distinction can be made between qualitative, comparative, and quantitative problems of value. The qualitative value-statement can be formulated that "X has value K," or "X is valuable (as such or instrumentally)." The comparative statement has the form, "X has more value than Y," or "X is more valuable than Y." The quantitative statement is, "X's degree of value is r." In qualitative statements, value can be classified by the qualitative labels, such as good, bad, and indifferent; high, medium, and low quality; excellent, very good, good, and fair.[16]

Statements	*General Form*	*Examples*
Qualitative (defining and classifying value)	X has value K X is valuable (as such or instrumentally)	Animals/living beings/ecosystems have intrinsic value
Comparative (ranking value)	X has more value K than Y X is more valuable than Y	Animals have more intrinsic value than plants
Quantitative (measuring value)	X's degree of value K is r	An animal's degree of intrinsic value is two times the intrinsic value of plants

The ranking and measuring of value challenges the idea of equal intrinsic value. Typically, we should not rank the possessors of intrinsic value. We do not say that Matti has more intrinsic value than Liisa. Could we say that Liisa's intrinsic value is at the level of fair or excellence? The idea of intrinsic value is that it is independent of merit. All who have intrinsic value have it equally. In this respect, Liisa and Matti are equal. This is the idea of humanity: all people are equal in virtue of their human inner worth. Do people have more value than animals or plants in virtue of their intrinsic value? The idea of equal intrinsic value is that the possessors of intrinsic value need to be protected. If the non-human nature has equal intrinsic value, we ought not destroy plants or ecosystems, or not kill animals for people's pleasure. To say that nature has intrinsic value is to say that nature has value for its own sake, as an end-value, in the language of intrinsic valuation.

C. Intrinsic Value as an End Value: Nature for Its Own Sake

According to the distinction of intrinsic and instrumental valuation, nature has intrinsic value if it has value for its own sake, and instrumental value if it has value for people. Gilbert H. Harman [17] refers to this kind of view as the standard theory of valuation accepted by John Hospers, William Frankena, and Richard B. Brandt. This standard theory of value in itself is not an anthropocentric value-theory. It does not say anything precise about the value content of things. It says that things can be valued intrinsically or instrumentally. Secondly, as a theory, the standard account does not claim that only human beings are valuers. People are not the only beings able to value things. The most intelligent animals, like chimpanzees, have this ability analogously with us. This ability of valuing things does not require an ability for conceptual or linguistic thinking. The gorilla Koko intrinsically valued her famous cat.

The distinction between intrinsic and instrumental valuation can be understood in terms of the value of ends and the value of means as the distinction originated in Plato and Aristotle. Some things are valued because they are means or instruments for some end, and this end is valued as worthy in itself. For example, we can value a forest as raw-material (means value) for human economic welfare (ends-value.) Intrinsic valuation is opposed to instrumental valuation. Applying this to non-human beings, we can ascribe to animals, plants, or ecosystems instrumental and intrinsic value. If nature conservationists value a wilderness area for its own sake, as an end, not deriving that value from instrumental grounds, then they regard it as intrinsically valuable.

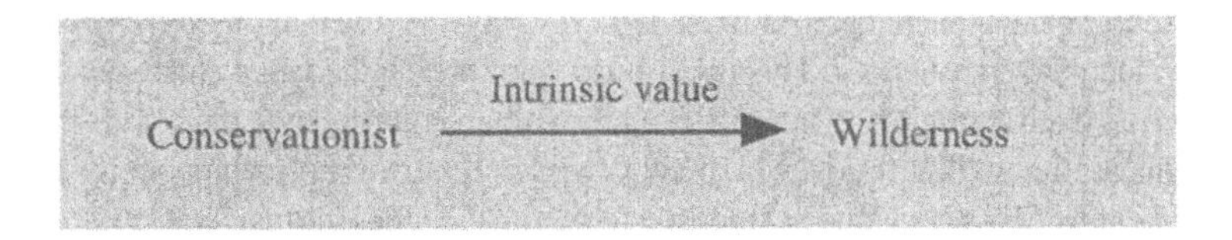

If some other value is derived from wilderness, for example, its raw-material value for the forest industry, then the wilderness area has instrumental value.

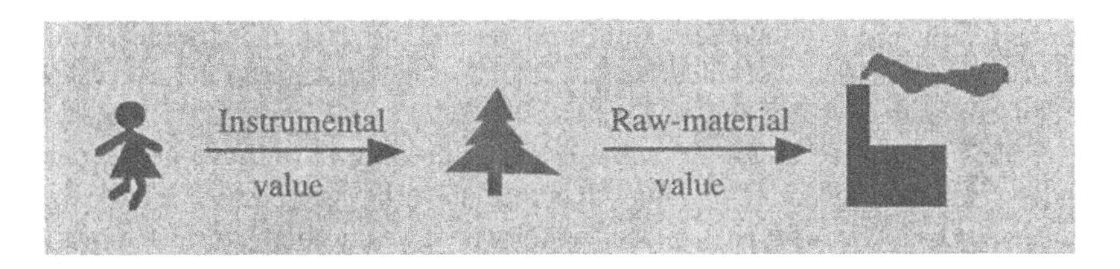

Wilderness can have different kinds of instrumental value, such as economic, aesthetic, scientific, historical, or religious value for people, and several use-values. Some instrumental value, such as recreational and experiential value can be compatible with the intrinsic value of wilderness. Animals and nature can be both intrinsically and instrumentally valuable, if we so desire. Natural beauty is a typical example of intrinsic valuation, valued for its own sake.

D. Intrinsic Value as an Inner or Internal Value: Nature Having Value in Itself

Charles Stevenson classified intrinsic and extrinsic valuation as follows:

> (1) "I approve of X intrinsically" has the meaning of I approve of X when I disregard all of its consequences upon other objects of my attitudes.
> (2) "I approve of X extrinsically" has the meaning of the consequences of X meet for the most part with my approval, and so I approve of X when I consider it with exclusive regard to its consequences.[18]

This distinction classifies "ends-values" in the class of intrinsic values and "means-values" in the class of extrinsic values. Can we accept the view that the object has instrumental value if it has no value for its own sake but only for the sake of the valuable things which are derived from it? If yes, value is given to an object because it has valuable consequences. This is extrinsic value which the value attributed to an object has, even though it is not intrinsically valuable for its own sake. It follows from this argument that things which are considered intrinsically valuable for their own sake have intrinsic value as opposed to extrinsic value. Intrinsic value is called "intrinsic" in the meaning that value is given, ascribed, or attributed to an object without references to the object's qualitative properties. This is J. Baird Callicott's sense of "truncated intrinsic value."[19] He truncates the old ontological sense of intrinsic. Robert Elliot has introduced the term "indexical" intrinsic value. For him, intrinsic value is a referred value. Both a referer and the referent are needed. This greatly resembles truncated value, because it "makes the existence of intrinsic value dependent on the existence of a valuer."[20]

This subjective theory of value is plausible only if we have no deeper understanding of the nature of values than that all entities get their value from the valuation-processes of people (or other beings who have the ability to value things). Things we value instrumentally as a means have extrinsic value, while things we value intrinsically as an end have intrinsic value. Meanwhile, nothing has value on its own, only as it enters into these relationships with valuing agents, human or otherwise. On one perspective, by saying that animals are intrinsically valuable, I attribute intrinsic value to the animals. On another perspective, I can mean that animals are intrinsically valuable in themselves. This suggests that animals have qualitative properties making them intrinsically valuable.

If the thing in itself is entirely valueless, without any value-making properties, how can I confer intrinsic value on it? The valuing process becomes arbitrary, unless based on some qualitative properties actually present. A possibility would be to introduce new terms, say, "internal" and "external," instead of "intrinsic" and "extrinsic." Or we can stipulate a better meaning for intrinsic, recapturing its old philosophical meaning. We should understand intrinsic and extrinsic values in a way that the terms "intrinsic" and "extrinsic" refer to the ontological states of values. The "in" (intrinsic) indicates the inner states of things while the "ex" (extrinsic) indicates properties or qualities which are not in the thing but outside it. Those values given or attributed to an object (without any reference to the object's qualitative properties) should be called the extrinsic or external value of the object. Values inhering in an object are the intrinsic, inner, or internal values of the object.

E. Intrinsic Value as a Qualitative Property of an Object

Some support for the intrinsic-value view of nature can be found in G. E. Moore. For him, intrinsic value is a non-natural property of an object. Goodness and beauty are intrinsic rather than extrinsic value in Moore's view. The goodness or beauty of an object depends solely on the intrinsic nature of the object in question: "To say that a kind of value is 'intrinsic' means merely that the question whether a thing possesses it, and in what degree it possesses it, depends solely on the intrinsic nature of the thing in question."[21] The problem with Moore was that he made a strict distinction between fact and value. He was unwilling to combine values into facts, because deriving values from facts was generally assumed to be the fallacy of the naturalistic theory of value. Thus, Moore claimed that some facts have "non-naturalistic" properties that make them valuable. Unfortunately, Moore was not able to define those non-naturalistic properties. My suggestion is to correct Moore's notion by saying that what makes some facts valuable is their qualitative, inherent properties: a qualitative property of a thing makes it valuable.

What are these qualitative, inherent properties? Three qualitative levels in nature are the level of consciousness, the level of life, and the level of nature as a whole. At the level of consciousness, reason and emotion are the examples of qualitative properties both in human beings and sentient animals. At the level of life, the qualitative, inherent property of a living being is its ability to preserve and reproduce life. At the level of nature as a whole, the qualitative property of nature is its inherent, complex life-support systems or its creative nature, nature as a whole at the ecosystemic level being able to generate varieties of facts and values in the world. Nature has ability to generate values, thus being valuable, not valueless, in itself.

F. Moral Standing

A distinction can be made between axiological (theory of value), deontological (theory of norms), and anthropological concepts (ethical issues related to the philosophy of human beings).[22] I add to this classification ethical issues related to philosophy of animals and nature that would be called "naturopology" in accordance with the term "anthropology." Or better, ethical issues related to nature form a part of nature-conservation philosophy.

Intrinsic values to	*for?*	*why?*
Human beings	Human beings God	Humanity Divinity
Nature	Human beings Nature	Humanity Naturity
Human beings Nature	God	Divinity

Nature-conservation philosophy is a three-dimensional responsibility: (1) responsibility for whom or for what, (2) to whom or to what, and (3) why so.[23] We have a sense of responsibility for our own sake or for the sake of nature. Our duty can be to ourselves, nature, or God. This can be because of prudential, religious, or ethical reasons. For example, all duties may be ascribable to God, not to human or non-human beings on the earth. We can ascribe intrinsic value to nature for the sake of human beings and because of humanity. This is the anthropocentric attitude to nature. Or if we intend to be properly naturocentric, we need nature-centered answers to these dimensions for, to, and why. Intrinsic values are ascribed to nature for the sake of nature because of "naturity."

Kenneth Goodpaster has distinguished moral considerability from moral significance. I have based my classification of moral standing on this distinction in the following, by adding the issue of moral priorities to it.[24] Saying that a being can be morally considered differs from saying that it should be morally considered to any particular degree. Moral significance refers to such things as should be taken into account in morality, whereas the concept of moral priority refers to comparative issues, for example, whether a plant should be taken into moral account more or less than other beings. One person asks whether animals or nature count at all (moral considerability), while another person asks whether they are morally important to be taken into account to any particular degree (moral significance). The third person may ask how much they count in comparison to other things (the issue of moral priorities). Should all of these things which can be taken into moral account be taken into account? Do they have moral significance? We can claim that nature is morally highly significant, as I think that human beings are. Yet my naturocentric claim is that people are automatically not superior to the non-human nature. The priority issue remains. We need to settle the dispute, roughly, of nature or people first. The claim that animals should be morally considered to a high degree, because they have highly sophisticated mental states being thus the subjects of their own life, does not solve the priority dispute in decision contexts, for example, are we allowed to kill and eat animals?

MORAL STANDING		
CONSIDERABILITY	SIGNIFICANCE	PRIORITY
Definitive: *Can* nature be morally considered? (Example: Plants have no interests; thus, we can do nothing morally wrong by killing them.)	*Normative*: *Should* X be morally considered to a particular (qualitative or quantitative) degree? (Plants have a good of their kind; thus, we should not kill them without good reasons.)	*Comparative*: Should the intrinsic value (or instrumental value) of animals have *priority* over the intrinsic value (or instrumental value) of plants (in a particular condition)? (An endangered plant species has priority over a common animal species.)

Ethical extensionism is an answer to the problems of moral considerability and moral significance. The basic argument is that different entities belong to our moral scope on different grounds: animals as sentient beings (zoocentrism), living organisms as having a good of their own (biocentrism), and the whole planet Earth on the grounds of its unique life-support systems (ecocentrism). Could it be that forests get their value only as being a home for sentient beings? Ethical extensionism finally accords intrinsic value to forests, rivers, mountains, which are thus to be respected and preserved. All of these three categories are morally highly significant. They should be morally considered to a high degree, although I have no simple practical answers to the priority issues between these categories.

Two

GOODNESS IN NATURE

Nature as good in itself should be distinguished from nature as good for people. My question is: what is good for nature? Intrinsic goodness is good in itself, independent of other things. All goodness is not intrinsic goodness. We need other forms of goodness. Two other forms of goodness are extrinsic goodness (subdivided into instrumental and transformative goodness) and systemic goodness. Systemic goodness refers to relational goodness. Good-in-itself is good in isolation; systemic good is good-in-relation or good within some other thing. With these distinctions, we are able to formulate the concepts of intrinsic goodness, extrinsic, and systemic goodness, which are basic forms of value. This classification originates in the philosophy of Robert S. Hartman (1910-1973).

1. Nature as Good in Itself

By saying that something is good in the instrumental sense, we mean that a thing is good of its kind by being good for some purposes. A good knife, as a knife, is good, if it is sharp. Sharpness is a good-making property for knives, in this instrumental sense. Attributing instrumental goodness to some thing is to say that this thing serves some purpose well. We must distinguish between something being a good kind of a thing for some purposes from something being a good of its own kind. Something being a good kind of thing means being good for some purposes relating to its instrumental goodness, as money is an instrumental value for human welfare. We may speak of a good tree or a good dog. Speaking of X as a good one of its kind refers to X having most of the features of a tree or a dog as a paradigm case. We need to distinguish "good for" from "good in itself," or, "good for" from what is "good." The practical conservation issue arises from this: should non-human beings be protected because they are good for us, or are they good in themselves? Animal activists claim that a sentient animal does not exist for anything else; it exists for itself. Being a sentient animal is a good thing. No human being is needed for the existence of animal valuers and their values: sentient animals value their own lives intrinsically. And they value the resources they use instrumentally.

2. Intrinsic, Extrinsic, and Systemic Goodness

Being good in itself, or, directly good, refers to intrinsic goodness, to be good as such. An object is good to the degree that it fulfills the conditions required by its own nature.[1] Something is good when it is what it is required to be by its indispensable nature. To fulfill itself is to be agreeable to itself. For example, if human nature requires an exercise of human reason, then a human being is good to the extent she or he exercises her or his reason.[2] A connection exists between the basic abilities of human beings and human well-being. We can reasonably assume that the nature of each living creature and its well-being are connected.

The postulation "nature is good" has three different perspectives; good in terms of a source of requirements (1) internal, (2) external, or (3) systemic to nature. If the source of requirements for value is derived from the internal specification of nature, we speak of intrinsic goodness. The essence of nature is a source of requirements for value. Saying that X is a good spruce tree refers to its intrinsic goodness. X answers the requirements of being a good spruce in its essential kind, as a spruce, to a degree that it would be unreasonable to demand more. This intrinsic goodness of a spruce must be distinguished from its instrumental goodness, for example, being a good spruce as timber for a forest industry.

The definitions for value from the internal, external, and relational points of view are as follows.

(1) Intrinsic value: by saying that A has intrinsic value, we are claiming that A satisfies its internal requirements for value.
(2) Extrinsic value: by saying that B has extrinsic value, we are claiming that B satisfies external requirements for value.
(3) Systemic value: by saying that C has systemic value, we are claiming that C satisfies functional or processive (relational) requirements for value.

By internal requirements I mean that A's intrinsicness, or its inner qualitative nature, defines its value. No outside reference is needed in defining A's intrinsic value. B has extrinsic value if its value is defined from the point of view of other than B itself. What is extrinsic for B, that is, external causes or factors, defines B's value. An example of extrinsic value is instrumental value, B being an instrument, a tool or raw-material for some other party's purposes. An example of systemic value is ecological value. The value of ecosystem is interconnected to the ecological functions of nature as a whole.

Intrinsic goodness can be explained in terms of essential value, and systemic goodness in terms of relational and functional values in nature as a whole. Instrumental goodness can be explained best by terms of tools and usefulnesses. While instrumental goodness exists between individuals (for instance, a deer has instrumental value for a wolf, who is hunting it), systemic goodness exists at the level of ecosystems (as an ideal case, wolves hunt unhealthy individuals, thereby sustaining the overall well-being of a deer population.)

	Human beings	*Non-Human beings*
Intrinsic goodness: X is good in itself	This is good as a wo/man	This is good as a spruce
Instrumental goodness: Y is good for X as a tool or a resource	This is a good wo/man for her/his children This is a good spruce for our bridge	This is a good spruce for birds This is a good wo/man because s/he feeds birds
Systemic goodness: X and Y are good in functional relations to each other	Man and woman are good in relations to each other Individuals and society are good in relations to each other	The members of a wolf pack are good in (functional/ecological) relations to each other Flowers and bees are good in (functional/ ecological) relations to each other The lynx and hares are good in (functional/ ecological) relations to each other

Animals and plants are ecologically connected to other beings and to the environment. From this does not follow that they could lose their essential value. Both kinds of goodness exist, in addition to instrumental goodness: (1) goodness in the essences of beings and (2) goodness in the necessary relationship of beings in nature. Intrinsic, instrumental, and systemic goodness are ontologically distinguished from each other, although they also are in some relation to each other. A tree's goodness requires the tree being in relation to other entities and to the environment. The totality of the value of the tree is not an isolated value inhering in the tree, although its intrinsic value depends on its inherent qualitative nature, that is, on its intrinsic goodness. Animals and plants cannot be separated from other beings and from their environment, and the same holds true of their values, which forms their functional or systemic value in nature.

3. What Is Good for Nature?

The biological view of the good of living beings originates in the statement that "a being, of whose good it is meaningful to talk, is one who can meaningfully be said to be well or ill, to thrive, to flourish, be happy or miserable."[3] All living beings have their own well-being. Each living creature has life, which can be destroyed or benefitted, and thus each living creature has its biological good. To have life means to have a biological telos, and to realize this telos can be said to be good for each living being. We human beings have an ability to take another being's standpoint. We can get empirical knowledge of the well-being of each living creature and take it morally into account. Thus, well-being is the fundamental biological and psychological basis for the intrinsic value of living being. The intrinsic value of living beings is analyzable in terms of the biological goodness of these beings. The opposite view stresses that the intrinsic value of an animal or a plant inheres in the animal or the plant, not in their states of well-being. To reduce the intrinsic value of animals to their good, or to their well-being, makes them valueless in themselves since their well-being is valuable, not the animals themselves. The mistake of this view is to distinguish the well-being of an animal from the animal itself. This distinction can be made conceptually but not ontologically; the well-being of an animal cannot exist in isolation from the animal. We should recognize that pain and pleasure can be understood in terms of the being as a whole. This point may be extended beyond the individual wholenesses of beings. We should not isolate individuals from their environments. The well-being of an animal or a plant requires that its environment also be well. This kind of ecological relation does not mean the reduction of the value of individuals to the value of their environments or the derivation of the value of individuals from that environment. It stresses the systemic value of nature, or the ecological necessity of the environment.

From the axiological point of view, the varieties of goodness in nature are related to what people ought to seek and promote. Goodness in nature should have moral standing. All non-human beings having goodness should count as morally relevant beings. Their goodness is a morally significant feature that obliges us. When we say that non-human beings possess axiological goodness, we do not mean that plants and animals are moral agents, but that their goodness ought to count in our moral concern. They are the objects of our morality. The basic question is, which kinds of goodness exist in nature? Some thinkers may hold that non-human beings possess biological goodness but not axiological goodness. This distinguishes between ontological and ethical issues. The ontological question, whether living beings have a good of their own, should be distinguished from the question, how much should their good count in human morality.

4. The Plural World of Value

My analysis above of goodness does not confirm a typical claim that the theory of values in nature would be developed best by explaining all natural values in terms of subjective or relative values.[4] A theory of value, which claims that all values are relational, is not plausible in thinking of the plural world of value. We should not reduce all natural values to the class of relative value or believe that all values can be explained in relational terms. From the literal point of view, we can accurately separate intrinsic, instrumental, and systemic value.

The varieties of goodness

- Intrinsic goodness (X being-good-by-its essential/inner nature; being-good-in-itself, X is good as what-it-is)
- Instrumental goodness (Y being-good-for X)
- Systemic goodness (X, Y, Z ... being-good-in their functional/ecological relations to each others)
- Biological goodness (P having-a-biological-well-being)
- Axiological goodness (P having moral standing on the grounds of its varieties of goodness)

From the substantial point of view, our approach should be to seek and classify all forms of value in nature. In addition to relative value, several other values exist. When a wolf eats a deer, no relative value from the deer's point of view is present. In this process, the deer has lost its life and its intrinsic value, although it still has instrumental value for the wolf, and it also can have systemic value (considering the broader life-and-death-roles in nature). Systemic goodness can be explained in terms of relational and functional value in nature, while I explain intrinsic goodness in terms of essential value, and instrumental goodness in terms of being an instrument, a tool or raw-material for other entities. These are three basic forms of value, which cannot be reduced to each other. To explain (natural) systemic value further, I recommend that "ecological" and "functional" are better terms than the term "relational," which is philosophically ambiguous. The term "relational" does not express the processive, functional characteristics of the ecosystemic level of life. We should not use the term "relative value," when we have the better term "systemic" value, which describes the functional and ecological characteristics of value in nature.

Evidently, Robert S. Hartman got it right, when he proposed this new form of value, systemic value, in addition to intrinsic and instrumental value. Systemic value can be distinguished both conceptually and ontologically from intrinsic and instrumental value. We do need these three forms of value to classify values in nature. We need both instrumental and systemic value to keep them separated from the question: what is the intrinsic value of nature?

5. Summary

Intrinsic goodness refers to the fact that the value of beings depends on the beings' own qualitative nature. The good of living beings refers to a biological meaning of good: the term "good" can be attributed to beings having life. Each living creature has life, which can be destroyed or benefitted, and thus each living creature has its biological good. Each living being has its essential nature, which it defends as a good thing (being-good-in-itself). Systemic goodness in nature refers to the fact that all beings are ecologically connected to other beings and to the environment. Yet their value can be isolated from other beings or from the environment: biological goodness is the goodness all living beings have because they can be harmed or benefitted in the sense no machine can. On the grounds of different kinds of goodness, we can argue that the goodness of non-human nature is evident, insofar as one can see its species, its niche, or its trophic level. The good and well-being of an animal is not obscure. A human being can distinguish whether an animal is suffering or not. As we make moral judgments of the well-being of other human beings, so we do of the well-being of animals. If we follow the biocentric thesis of the good of each living creature, we can also distinguish whether their non-human nature is damaged or not, whether there is ill-being or well-being. If we consider nature from the intrinsic-value point of view, to be a good of its kind does not mean instrumental goodness or being good for some purposes, but being good in what it is. Intrinsic goodness refers to things as what they are by their inner qualitative nature. The systemic theory of value relates nature's value to the functions of nature as good in some respect, while intrinsic-value theories relate nature's value to the essence of nature as good in itself. Both types of goodness can exist in human beings and in non-human beings. The third basic type of goodness is instrumental goodness. Something is instrumentally good when it is an instrument or a tool for another entity.

Three

THE FORMS OF INTRINSIC VALUE

1. Four Groups of Value

In this chapter, I classify the basic meanings of intrinsic value into four main groups: (1) Lewis's intrinsic value, (2) values according to Aristotle and Kant, (3) weak intrinsic values, and (4) strong intrinsic values. These different meanings of value answer the question, "What can be meant by the concept of intrinsic value?", instead of the question, "What kinds of things have intrinsic value?", which is discussed at the end of this chapter.

1. Lewis's intrinsic value
A. experienced or sensed value
B. immediate value
C. non-derivative value

2. Values according to Aristotle and Kant
A. ends-value
B. self-value

3. Weak intrinsic values
A. uniqueness value
B. inherent value

4. Strong intrinsic values
A. inherent worth
B. inner value
C. essential value
D. existence value
E. inherited value

In Group 1, pleasure is found as directly good. It is immediately and intrinsically valuable in life.[1] "Satisfying experiences, pleasant experiences, happy experiences, all hedonically toned states of consciousness" are intrinsically good.[2] The source of intrinsic value is consciousness, all kinds of experiences: "If an experience, or anything else, has intrinsic value this value belongs to it in virtue of its own nature and is independent of any derivative value it may possess."[3]

Pleasure is typically claimed intrinsically good, while displeasure i s said to be intrinsically bad. Someone experiencing pleasure is intrinsically good, while someone being conscious is intrinsically good. Intrinsic value in terms of experienced or sensed value was established by that eminent philosopher C. I. Lewis and adopted and elaborated by many philosophers. Both immediate and non-derivative value are close to experienced or sensed value. Immediate value can be characterized as indicating the value perceived instantly, intuitively, or instinctively. What is value in itself and not in its consequences, is a non-derivative value. A modern defense of intrinsic value in terms of experienced-value is Michael Zimmerman's intrinsic-value theory of the states of pleasure.[4] Archie J. Bahm distinguishes between three basic kinds of intrinsic value, namely pleasure, satisfaction, and preferences, all human feelings.[5] Eero Paloheimo has also defended intrinsic value in terms of experienced states. All sentient animals have those experienced states, thereby being intrinsically valuable beings. For Paloheimo, a sentient animal is *Coccinellidae*, the ladybug. This is reasoned by arguing that ladybugs have sense organs, which makes them the experiencing subjects.[6] Paloheimo's view is original. The anthropocentric claim would be that human beings can have intrinsic value in terms of states of experience, while the "mammocentric" claim would be that some well-developed and well-trained mammals have this kind of intrinsic value.

As to Group 2, which includes ends-value and self-value, I have derived the standard of intrinsic value from Aristotle and Kant, even though they did not use the term "intrinsic value" at all. The idea of an object as good for its own sake originates in a well-known Aristotelian division: something being an end to be pursued for its own sake as opposed to something being a means pursued for the sake of some end. So something has intrinsic value if it has value for its own sake, but instrumental value if it has value for something else.[7] Intrinsic goods are things desirable for their own sake, or as ends in themselves, inherently worth, to be actualized and preserved.[8] Kant argued that animals are merely a means to an end, and this end is humanity.[9] Saint Thomas Aquinas in turn claimed that only the intellectual nature of human beings exists for its own sake, all other things exist for the sake of human beings.[10] These kinds of anthropocentric views have undoubtedly dominated Western attitudes toward nature; nature and non-human beings have been seen merely as means to human ends. This is a historical fact about value, not a philosophical necessity. In contemporary Western societies, animals, plants, and other non-human entities are evidently valued for their own sake by many people.

In the classification, intrinsic value in terms of self-value has two interpretations: (1) the old Kantian view of self-estimation, and (2) the zoocentric view that represents animals as subjects of their own lives. Kant expressed human value by using terms "inner worth" and "dignity." People in their highest self-esteem are above all price.[11] The idea is similar to ends-value among human beings. People should not be valued only as a means to the ends of other people, but they should be prized as ends in themselves. They possess a dignity or an absolute inner worth. People, who realize their own intrinsic

Meanings of intrinsic value	Characteristics	Authors
Experienced or sensed value	pleasure/interests/ conscious states as good	Lewis, Perry, Zimmerman, Singer, Paloheimo
Immediate value	instantly/intuitively/ instinctively grasped good	Moore
Non-derivative value	good in itself and not in its cause or consequences	Baylis, deontologists
Ends-value	good for its own sake	Aristotle
Self-value	1. The highest appreciation of the self	Kant
	2. The subject of its own life	Regan
Uniqueness value	1. Be yourself! 2. Value of species	Rousseau Nature preservationists
Inherent value	good by its own nature (art)	Taylor
Inherent worth	a good of one's own (living beings)	Taylor
Essential value	good of a kind	von Wright
Existence value	goodness in beingness	Spinoza
Inherited value	biological value, survival value	Biologists

value, esteem all other people within this value. We human beings have intrinsic value in terms of our ability to value ourselves. I have ability to value my own life, which demands that I should respect the others who also have this ability. The Kantian-like argument runs: Intrinsic value refers to the idea that we have ability to value ourselves. I have an ability to value different kinds of states and things. Additionally, I have an ability to value myself. Therefore, I have intrinsic value. This argumentation can be used in the case of animals in that animals have ability to value themselves, and thus, have intrinsic value in terms of the "experienced-value" of Lewis and the Kantian "self-value."

In Group 3, the inherent value of an object means that its value depends on its nature and not its consequences. Inherent value is defined as differing from sensed value with respect to the object which has this value. Sensed value is located in conscious states; inherent value is located in objects outside conscious states. Those things we want to preserve as such, independent of their possible derivative value, which therefore have inherent value, can be listed as follows: works of art, historical buildings or places, battlefields, wonders of nature, and so on.[12] Living things, like pets, can belong to this class of value. This is a weak objective value, because it depends on someone's valuing it. Inherent worth belongs to Group 4, the group of strong objective values. The idea of inherent worth indicates that if an entity has a good of its own, and if it is better that this good be realized than not, then the entity has inherent worth. For a living entity, it is better that its good be realized than not realized.[13] This refers to the value independent of valuers. Inherent worth is separated from values of any other beings and from merits. Those who have inherent worth have it in their own nature, and it belongs to them inherently.

The case of intrinsic value in terms of the value of existence considers that mere being is good.[14] In the case of essential value, nature's intrinsic values can be understood in terms of the value of the essences of natural beings. The essential value of a being refers to something in its own nature. It refers to a paradigm case, an instance having all or most of the features, and in a high degree, of a kind. Inherited value refers to such biological values as survival value. Intrinsic value in its strongest sense is the mixture of inherited, essential, and inner values; the intrinsic value of an object refers to a value which is inner or inside the object.

2. Intrinsic Value in Terms of Respect, Love, and Preservation

Classic absolute values are truth, goodness, beauty, and the value of sanctity, all still commonly approved in Western culture. The truth in terms of value is expressed as follows: "To see what a thing is, *whatever* it be, would mean the same as to see its value."[15] The value of love involves the ideals of sympathy and identification. Robert Hartman has elegantly said that the language of intrinsic valuation is the language of lovers, artists, mystics, and prophets. Those persons are willing to give their souls to their works. They become totally involved in what they love and appreciate.[16] The same kind of identification within nature is in deep ecology: love the animals, the plants, and the nature as a whole. The land ethicist, Aldo Leopold claimed that land should be loved and respected, which demand is a basic idea of the extension of ethics.[17] The value of love and the value of sanctity are interconnected in Henryk Skolimowski's philosophy, too.[18] Does the language of respect coincide with those intrinsic-value views developed by recent philosophers of animal and

Classic intrinsic values	*Characteristics*	*Authors*
Love value	identification sympathy	Hartman, deep ecologists
Truth value	how things are in themselves, to be what it is	(Naive) realists in science
Value of beauty	harmony, creativity	Hargrove
Value of sanctity	good such as God created	Schweitzer, Skolimowski

nature? Tom Regan's key concept of inherent value is related to an attitude of respect and to a preservation principle. Something having inherent value expressly means that it should be respected and preserved. The inherent value of a natural object is such that the fitting attitude toward it is of admiring respect, that is, the admiring respect of what is inherently valuable in nature gives rise to the preservation principle.[19] To analyze the notion of intrinsic value in terms of respect, we should also consider what Paul W. Taylor, author of *Respect for Nature*, says. The concept of inherent worth is a key part of Taylor's biocentrism. Biocentrism regards respect for living nature in the same way as anthropocentrism regards respect for human beings. All animals and plants have a good of their own. This means that for a good which is not derived from any other's good, no reference is made to other beings, whether conscious or non-conscious. Taylor's assertion is that every individual animal and plant has its own good. And this good of living beings and their inherent worth is of fundamental value in an attitude which encourages respect for nature.

Finally, I characterize the notion of intrinsic value in terms of uniqueness. The idea of uniqueness is closely related to the idea of respect for human beings in terms of intrinsic value, that is, the intrinsic value of human beings is the command to respect the uniqueness of each individual human being. Is this notion of intrinsic value applicable to non-humans? We say of our best friends that they are irreplaceable, priceless, inestimable beings. Kant's well-known claim was that in addition to our good friends, each human being should be regarded as priceless and as beyond any instrumental value. Could some animals be our irreplaceable friends? Several animal companions do have intrinsic value in this sense. And the eminent defender of the wolf, R. F. Dubois calls wolves his friends.[20] Could animal lovers extend their concern of animals to all sentient animals? Peter Singer and Tom Regan, among others, have attempted to justify this extension. Singer's well-known claim is that you need not be the friend of an animal to want that it should be treated with justice.[21] How about non-sentient creatures? Have they intrinsic value in terms of uniqueness? The notion of the diversity of life is one of the most powerful recent ideas of nature preservation. And the notion of diversity is closely related to the idea of uniqueness of life forms. The old principle of the plenitude suggested the commandment: Be yourself, be unique![22] This stresses the value of being original, which is related to the idea of the expansive process of life: the whole world manifested itself in the maximal differentiation of the creatures. These old notions primarily concerned human uniqueness. My point is that each species has a unique value, for each species is irreplaceable. No sum of money could replace the value lost by the extinction of a species.

A crucial notion of intrinsic value is in terms of respect. To say that X is intrinsic value or has intrinsic value, is to say that X should be respected, for the sake of X's uniqueness. This is what can be meant by human intrinsic value, and it is quite applicable to animals and nature. The language of intrinsic value is the language of respect and preservation. It is opposed to the view which regards treating plants and animals as objects to be manipulated rather than respected and preserved. Nature having intrinsic value means that it deserves respectful treatment. Objects having sole instrumental value do not deserve respectful treatment such as they are, since they may be used without qualification as instruments for others. This distinguishes pure instruments from things with both intrinsic and instrumental value (like human beings, animals, living organisms, and ecosystems). This is not to say that the instrumental-value view by necessity leads to the destruction of the non-human environment. People can preserve nature for the sake of its instrumental value or transformative value. Many varieties of values in nature should be respected. Furthermore, all tools and good instruments should be treated appropriately.

3. The Intrinsic Value of Animals, Plants, and Ecosystems

I question the ethical and ontological view, that the non-human world has solely anthropogenic and anthropocentric value. Any criticism needs a more positive answer to establish what values are beyond anthropogenic and anthropocentric values. From the ethical perspective, I develop the naturocentric argumentation in which different kinds of entities belong to the scope of our morality on different grounds: animals as sentient beings (zoocentrism); living beings because of the value of life (biocentrism); the whole planet Earth because of its unique life-support systems (ecocentrism). From the ontological point of view, I develop an answer in terms of naturogenic value: the value which a non-human entity has because of its inherent qualitative characteristics. Naturogenic values are intrinsic, qualitative properties in nature. They are generated by animals (zoogenic value), by living beings (biogenic value), and by ecosystems (ecogenic value). Animals, plants, and ecosystems are able to generate values, thus being value-laden and not intrinsically barren of values.

The Intrinsic Value of Nature	
Naturocentrism (zoocentric, biocentric, ecocentric value)	Naturogenism (zoogenic, biogenic, ecogenic value)
Value centered on nature, the issue of human place, and relations to nature	Value rooted in nature, the issue of the origins of value
The ethics of value	The ontology of value

My attempt is to show that developing the topic of naturocentric and naturogenic value is worthwhile.

PART II

NATURE CONCERNS

Four

ZOOCENTRISM

A methodological tool to organize the problems that surround the issue of intrinsic value is ethical extensionism. It extends the scope of ethics from people to animals through life-forms to ecosystems. In this book, the definitions for zoocentrism, biocentrism, and ecocentrism are as follows:

(1) Zoocentrism means a philosophy in which the issues, concepts, and values of animals are central.
(2) Biocentrism means a philosophy in which the issues, concepts, and values of life are central.
(3) Ecocentrism means a philosophy in which the issues, concepts, and values of ecosystems are central.

Zoocentrism covers the discussions in which the notions of higher animals and their value are central. Zoocentrism is the animal-centered, especially vertebrate-centered philosophy. The point of view is animal preservation, rooted in animal rights activism. This differs from the biological point of view in that the main concern is animal suffering in human societies (concerning laboratory animals, meat animals, trade animals, and fur animals). The concept "biocentric"

Naturocentrism	*Zoocentrism*	*Biocentrism*	*Ecocentrism*
A key concept	Consciousness	Livingness	Wholeness
Value	Well-being	Biodiversity	Beauty
Field	Animals	Life	Ecosystems

is sometimes restricted to mean Taylor's biocentrism. I define biocentric to cover the discussions in which the notions of life and living organisms, including lower animals, are central. By ecocentrism I mean the discussions of ecosystems and of their value. I discuss these naturocentric philosophies in the following chapters. I begin with zoocentrism.

What are animals and what is their welfare? The idea of animal well-being is a means of countering speciesism. The core of Western morality has been the idea of human well-being and welfare, the good of people. Human life has been supposed to be sacred and to have a special importance that no other animals have. The central concern has been the protection and care of people, while other animals do not have the same degree of moral protection, or else they entirely lack moral standing. Even animal sciences have been focused on human welfare or health. Animals have been treated as instruments and models for human illness or as raw-material for the human economy (meat and fish sciences, fur and wool economics).

Some philosophers have argued that we do not need to postulate intrinsic value for animals to understand the importance of their welfare. One possibility is to accumulate diverse arguments and find that the notion of intrinsic value is also important for the conception of animal welfare. I would defend pluralism in which many reasons exist for liberating animals. Liberating animals changes the world into a morally better place by developing the human moral character. It makes the world a fairer place, and a happier place for animals as well as for people.[1]

The modern animal rights movement is based on the assumption of the intrinsic value of individual animals. It fights against the instrumental model of animals which sees animals as resources and raw-materials for industrial, commercial production, and consumption. This view is manifested in factory farms in which animals are treated as masses and machines, not as individual beings and as the animals they are. Nowadays those who perceive animals as sentient beings cannot accept the treatment of them as raw-material for human usage. They are obliged to treat animals from the point of view of animal well-being by respecting them instead of utilizing them for human purposes. The practical situation of animal welfare seems hopeless. The mass of farm animals are treated more or less cruelly.

This chapter aims at dealing more extensively with the well-being of animals. What is good for animals? The basic issues of animal welfare are (1) animal suffering and well-being, (2) animal subjectivity, (3) animals as sentient beings, and (4) animal rights. Animal rights and other philosophical questions about human relations with animals have been massively studied recently (for a list of basic authors, see Chapter Nine). According to a zoocentric ethics, animals belong to the scope of our morality on the grounds of their sentient nature. This zoocentric view is a human perspective, although aimed at taking seriously the welfare of animals. It demands being humane and not brutal to animals for the sake of the well-being of animals themselves. Zoocentric ethics is not an anthropocentric attitude because it is not reasoned or grounded from the perspective of human well-being. Zoocentric interests are human interests in the non-human world. They are based on the well-being of animals, and not primarily on human well-being. An animal's ability to suffer and feel pain and its psychic stress, as well as its ability to value its own life should be taken into moral account. Animals should be valued in terms of their well-being, and this well-being should be determined from the animals' point of view. This defends the intrinsic-value view of animals, for which the opposing assumption is to see animals as instruments for human purposes.

1. Animal Suffering and Well-Being

The discussion of animal welfare has recently been focused on the question of suffering and pain. If suffering has moral significance, then all entities which can suffer are morally considerable and must be taken into moral account. Thus, all sentient animals belong to our moral community in the sense that we are morally not allowed to have an indifferent attitude to animal suffering. But what is this animal suffering or pain? In the following, I introduce some foundations from the field of animal sciences, underscoring the significance of empirical studies for the philosophy of animals.

Animals can suffer from a lack of control and from a disturbed relationship with the environment. Three basic indicators exist for assessing suffering in animals: (1) physical health, (2) physiological signs, and (3) behavior.[2] Physical health is a criterion for assessing a part of animal well-being. The signs of injury or disease are symptoms of illness, while bright eyes, sleek coat or feathers can be the signs of health. The problem is that physical pain behavior is not a direct indicator of suffering. J. S. Kennedy has argued that even the ascription of pain to animals is a new anthropomorphism, the ascription of human mental experiences to animals.[3] I object to Kennedy's position by arguing that finding something which animals share with human beings, as capability of feeling pain, does not make them analogous with us any more than making us analogous with them. Epistemologically, we can learn of their pain by comparing it with our experience and behavior, although their pain is real pain, actual pain, as is ours, not something analogous to our pain. The assessment of the physiological signs of suffering in animals means monitoring physiological processes, such as changes in hormone levels, brain activity, heart rate, and body temperature. The whole complex of physical changes is called (physical) stress. The behavior of an animal can be an indicator of mental suffering, suggesting its mental states, for example, that the animal is frustrated or afraid. A method of "asking animals" has been developed by animal-welfare scientists. One proposal is an integrated approach which utilizes different methods in assessing animal suffering. An animal's good welfare can be recognized by quantifying animal preferences and signs of pleasure. The integrated approach to welfare includes assessment of health and the appearance of well-being, productivity, physiological and biochemical parameters, assessment of behavior (normal, abnormal, altered), and the recognition of pain and suffering as expressed by different species.[4] The basic needs of animals are as follows: freedom to perform natural physical movement, association with other animals (appropriate to their own kind), facilities for comfort-activities (rest, sleep, bodily care), provision of food and water, the ability to perform daily routines of natural activities, the opportunity for the activities of exploration and play (especially for young animals), and the satisfaction of spatial and territorial requirements (a visual field and individual space).[5] Complete animal welfare involves all of these requirements.

The productivity of animals (rate of growth, milk production, and egg yield) is an aspect of animal well-being. Diseases and injuries indicate physical illness. But we would be mistaken to suppose that this is the whole truth of animal welfare. Productivity is in the interest of farmers; healthy functioning is in the interest of animals. Physical illness means decreased well-being, although during recovery, well-being may be increasing. The lack of illness does not necessary indicate the state of well-being. Animals can express psychological stress, although they do not have the physical signs of stress. The health of an animal can be decreased even with no physical sign of illness. The lack of illness can be said to be a necessary condition for animal well-being, although it is not a sufficient condition.

The concept of human health is extensively studied in the philosophy of medicine.[6] The World Health Organization (WHO) has defined health as a state of mental, physical, and social well-being. I propose that this definition is a good guide for animal welfare. Human animals as well as non-human animals have several psychological (mind), physical (body), social, and environmental requirements for their welfare. We need animal subjectivity to understand what is good for animals.

2. Animal Subjectivity

Are human beings the only beings who look at the world from their subjective point of view? Although we cannot know exactly what it is to be an animal, say a cow, we know that the cow has a well-being of its own which definitely is not human well-being. We human beings have the ability to understand something of the cow's viewpoint. The philosophy of animal mentalism stresses the private internal states of animals.[7] The question of "What is it like to be a bat or a cow?" is central. The general requirement for the subjective character of experience means that "there is something that it is like to be the organism – something it is like *for* the organism," to use Thomas Nagel's words.[8]

The subjectivity of animals has been investigated by Donald Broom, Marian Stamp Dawkins, and Françoise Wemelsfelder. They attend to the significance of the environment for animal well-being. Dawkins has even argued that the ultimate goal of animal welfare studies is the subjective experiences of animals.[9] People can have genuine empathy with animals, with how things are from the animals' point of view. We understand something of the reality of a squirrel.[10] We should begin to see the world through the animals' eyes.[11] Asking animals themselves is getting the animals' opinion. The method of operant conditioning means that an animal has the chance to learn; for example, that by pressing a lever it gets food (a reward) or can avoid something (a punishment).

Animal consciousness and suffering exist on the cognitive and emotional levels. Vertebrates deal with their environment in an emotional way. They can suffer pain and experience pleasure and well-being. The ability of operant learning means that animals are able (more or less reliably) to control how to obtain food or escape from danger. An animal detects order in its environment, and its brain does so in an emotional way. An individual obtains information from its environment. Its brain is a cognitive map neurally representing the environment which guides it in searching for food, shelter, or social contacts. Welfare includes physical and mental health, and mental health is strongly linked to this brain-behavior relationship. Highly monotonous conditions mean that individuals cannot actively collect information which may affect their well-being. Long-term uncertainty or loss of control may cause chronic stress in animals.[12] The typical view of animal welfare sees control as central for animal consciousness. The alternative view is based on an animal's subjective relationship with the environment. Even simple learning behavior can be based on an animal's subjective awareness of the environment. Subjective feelings can be supposed to have evolved because they help animals to avoid death or failure to reproduce.[13]

Animal-welfare sciences are not value-free. These sciences are able to solve the problems of animal welfare by taking ethical issues seriously. The fundamental questions of animal welfare are: What is animal welfare? What is the well-being of an individual animal? Why is it of importance? What is the ethical significance of animal well-being? These are moral questions. Health in human and animal medicine is a value-laden concept.[14] Although the questions of health and welfare (What is health? What is welfare?) are philosophical and ethical problems, this does not mean that animal welfare cannot be measured empirically. A prominent study is Wemelsfelder's *Animal Boredom*, which takes seriously philosophical and moral bases for the empirical study of animal welfare.

3. Animals as Sentient Beings

Michael Leahy detects two separate groups in the discussions about animal welfare. The first group is concerned with the question of animal welfare, concerning such concepts as being healthy and free from pain. Second group is concerned with animals as human beings. They have the human paradigm in mind when they attribute to animals such features as being happy or free from worry.[15] Leahy understands animal welfare in terms of the instrumental values of animals. Animals as pets, as a food source, are seen as tools or raw-material for us. This understanding of animals is inadequate. We find several other ways of perceiving animals that fall between seeing animals as raw-material for us or animals as human beings. I agree that animal welfare in Western societies is mostly defined by the standards of veterinarians, farm inspectors, animal managers, and so on. The terminology applied to animals and their welfare is not value-free. Applied animal sciences, such as meat science and milk science, investigate the question of animal welfare from the perspective of how animals are useful for us, not from the perspective of what animals are in themselves. If we respect the existence of reason and intelligence in the world, then we should respect animal reason as well as human reason. Darwin's theory implies that animals not only can suffer pain but are in many respects intelligent and sensitive beings. For Darwin, even worms, although they stand low in the scale of mental power, possess some degree of intelligence. He developed a project on the mental powers of worms. We should stress similarities between non-human animals and human beings: we are a great work that was created from animals.[16]

Leahy's suggestion that animals are primitive beings is nearer the view of animals as sentient beings than raw-material, but the interest of animals is not to be primitive creatures adapted to human life. The aim of other species aim is not to be adapted to human ways in any sense, but some species such as meat animals and milk animals are coerced into adaption to human ends. A cow's interest is not in being a milk-machine, nor is a pig's interest in being a meat-machine. From their point of view, the optimal life of a cow and a pig would be to be treated as intrinsically valuable beings, to live free, not to be consumed in any way, but to get enough food and shelter from people.

What is wrong with understanding animals as primitive beings? We misunderstand animals by assuming that being human is the end of evolution toward which all other species tend. Human beings are animals. We are one species of animals, like chimpanzees and dogs, but differences between chimpanzees and dogs are greater than the differences between human beings and chimpanzees. This is true morphologically if not culturally. Yet chimpanzees and dogs have genuine abilities that no human beings ever attain. Thus, considering animals as primitive beings and attributing pre-human abilities to them does not admit their distinct nature and unique abilities. Animals are not incapable sub-human beings with proto-human behavior and nature. Different kinds of animal abilities exist, although all of these may not have been found by research. More research which respects animals as sentient and individual beings is required to find and understand these abilities. In the following sections I consider Singer's view on animals as sentient beings, the problem of interests, and finally, Regan's view on the inherent value of animals.

4. Peter Singer, the Defender of Sentient Beings

In Singer's view, states of sentient beings are of intrinsic value and belong in the moral sphere.[17] He explicitly denies the possibility of intrinsic value and moral significance in non-conscious states, but defends the well-being of sentient animals. He objects that a plant could have well-being because intrinsic value is connected to conscious states. Can we compare the values of different conscious lives? If we compare different lives in a hierarchical way, as Singer suggests, we people inevitably place ourselves at the top, and the hierarchy of values remains anthropocentric. Singer asks us to imagine that we have the ability to turn ourselves into an animal's mind. In this process, I really am a horse and I have all the mental experiences the horse has. Further, I have the ability to enter a state in which I can remember what it was like to be a horse and what it was like to be a human being. I could then compare the value of a horse to the horse and the value of a human being to the human being, a scenario in which we could arrange our value-hierarchies from the point of view of another's life. A lot of difficulties beset this act of imagination. Thus, Singer prefers the view that each conscious life has equal value. We cannot judge our human lives as more valuable from our points of view than the lives of mice from the points of view of the mice.

Singer is right in that the conscious states of sentient beings have intrinsic value, but his mistake is to assume that all intrinsic values are necessarily related to conscious states. Instead, varieties of intrinsic values occur in nature in conscious and natural states, such as well-being, biodiversity, and beauty in nature. If nothing else but human life has value, then values would be a special kind of human property. Singer's argument is that nothing else but sentient life has intrinsic value.[18] If so, intrinsic values would be odd kinds of entities referring exclusively to the experiences of conscious states. To the contrary, intrinsic valuation can be applied to individuals, groups, and things. The world in which only the experiences of conscious states could be said to have intrinsic value would be quite different from our real world in which we find intrinsic values in many different kinds of things and states, not the least in natural states. As Mary Warren has maintained, sentience is a sufficient but not necessary criterion for the possession of intrinsic value.[19] The conscious states of each sentient being constitute an intrinsic value, but this is not the only intrinsic value in nature.

5. The Problem of Interests

Can intrinsic value be adequately analyzed in terms of interests? To whom can we ascribe interests? According to the zoocentric model of interest, the moral standing of animals is based on the view that animals have their own interests. Lawrence Johnson's original view is that non-sentient and even non-living entities have interests and thus intrinsic value. Thus, the relation of intrinsic values to interests is not necessarily anthropocentric or zoocentric as in Singer's view.

Animals have interests and thus belong within the scope of morality, or so I claim. However, the view of animals having interests is strongly questioned.[20] Two eminent authors of interests define interests in terms of human reason.[21] According to Paul Ziff, the disputes about meanings of intrinsically good indicate the differences of people's interests. He has analyzed different meanings of the word "good" concluding that "good," means answering to identifiable human interests. In his extensive study of interests, R. B. Perry rejects biological and even psychological views of interests, and defends interests as societal phenomena. Interests consist of the unity and totality of human life. Interests as an intrinsic value is normally an anthropocentric view, only human beings having interests are taken into moral account.

Perry has defined interests in terms of unities of human life; Singer defines interests in terms of experienced states. Typically, except for few biocentric thinkers, speaking of the interests of plants is regarded as nonsense, but the interests of animals may have some common support. The term "interests," at this stage of dispute over the many meanings of interests, seems too vague and contentious to be the key term with which we could locate intrinsic value. The term "biological interests" should be developed further, to distinguish biological interests from human or conscious interests.

6. Tom Regan's Theory of the Inherent Value of Animals

Regan posits two criteria for environmental ethics: ethical extensionism and the inherent-value view of nature.[22] He extends the scope of ethics from people to animals, and even to non-conscious beings. This enlargement is based on a value those beings have in themselves. In Regan's view, the inherent value of a natural object is independent of any awareness or conscious states. First, the value inheres in the object. This is an objective view of values. Second, Regan suggests that the presence of inherent value in the object is a consequence of some other properties the object possesses. Third, inherent value is related to an attitude of respect and to a preservation principle: if something has inherent value it should be respected and preserved. Although Regan does not claim that these points are established beyond dispute, he believes that they show the direction in which environmental ethics must move. Regan's later view is more sophisticated, but it retains the objective nature of inherent value described above.[23] He approaches the question of inherent value by discussing what kinds of entities have inherent value (those who satisfy the subject-of-a-life criterion) and what is inherent rather than intrinsic value (experiences, pleasure, and other mental states). He starts from the inherent value of moral agents: if moral agents have inherent value, so do moral patients.

In Schweitzer's reverence-for-life attitude, all living forms have their own value.[24] Regan does not agree with Schweitzer, because he is not sure whether being alive is a necessary or a sufficient condition for something having inherent value.[25] Regan rejects the being-alive criterion and suggests the subject-of-a-life criterion, which is the most important requirement for inherent value. Not all living beings satisfy the subject-of-a-life criterion, nor do all moral patients. But those who satisfy that criterion have inherent value. Since Regan rejects the view that inherent value could be something measurable they have it equally. He rejects the view that inherent value is to be linked to the mental states of individuals. He calls the value of mental states "intrinsic value," which is clearly distinguished from "inherent value." He claims that those who have inherent value are more than receptacles. This is clarified by the cup analogy. The cup itself is valuable, not just what goes into it, while on the receptacle view what goes into the cup would be valuable. Individuals themselves have value, not only their pleasures or preference-satisfactions.[26] For Regan, a contentless receptacle could be valuable. This leads to the distinction between the value of individuals themselves and the value of experiences they have. Inherent value cannot be reduced to the value of their experiences. Hence, Regan cannot accept that individuals are receptacles whose experiences are the only things that matter. He argues an alternative view, the view according to which individuals themselves have inherent value. The opposite view is that some states of beings are of intrinsic value, and not individuals in themselves. I suggest that both views are partly right. Individuals in themselves and their states of well-being have intrinsic value.

Ascribing inherent value to the individuals who are members of some group raises the question why just this group merely on the strength of their membership merit inherent value. Regan claims that demarcating human beings from non-human beings is arbitrary in the ethical framework.[27] We have no ground to the claim that all human beings have inherent value while no non-human beings have such value. To be human varies so much that we cannot find any property or ability that belongs to every human being but to no non-human being. We can postulate a morally significant separation between moral agents (most human beings) and moral patients (both human and non-human beings). This separation does not differentiate those who have inherent value from those who have not. On the contrary, Regan holds that both groups have inherent value. If this is true, then we have no grounds for restricting inherent value only to people.

Moral agents are those who can act morally. Moral patients are the objects of those actions. The paradigm of a moral agent is a normal adult human being. Moral agents have moral responsibility, intentionality, rationality, and so on, which are abilities moral patients may lack. This separation is the same as Taylor's concept of moral agents and moral subjects.[28] Moral agents are those who can treat others rightly or wrongly, and moral subjects are those treated rightly or wrongly by agents. An intrinsic link can be found between inherent worth and moral subjects. If an entity has inherent worth, then the entity is a moral subject which deserves moral concern. The corresponding view in Regan's theory is the attitude of respect for moral patients. According to this attitude, doing them harm matters to moral patients. Moral patients can be treated rightly or wrongly, and moral agents have a direct duty not to harm another, whether it is a moral agent or moral patient. If inherent value can be postulated for moral agents and moral patients, the question arises: who are those moral patients? Regan distinguishes two groups of moral patients:

(a) those individuals who are conscious and sentient (or can experience pleasure and pain) but who lack other mental abilities
(b) those individuals who are conscious, sentient, and possess the other cognitive and volitional abilities.[29]

Most mammals belong to group (b), while other animals to group (a). Regan has criticized a categorical criterion which demands that only individuals of human species are moral patients and have inherent value. Regan himself is claiming that only animal species are moral patients.[30] This is a criticism stated by some ecofeminists and ecoholists, who argue that ethical extensionism which restricts morality to animals, perhaps only to some mammals, is no new environmental ethics at all but a kind of sentient humanism, or sentientism. It is an ethics for people and their near cousins.[31]

Jan Narveson has criticized Regan's view.[32] He claims that positing of inherent value in all moral patients is invalid. Regan argues that (1) if moral agents have inherent value, (2) we have duty not to harm them in virtue of that value, (3) if some harms done to them are identical with harms done to moral patients, and (4) if we have a direct duty not to harm either moral agents or patients, then it follows that (5) moral patients have inherent value. Narveson is right that this argument (1) to (5) is not logically valid. It need not be the case that moral patients must have inherent value if they can be harmed in the same way as moral agents which have inherent value. Possibly the inherent value is so linked with moral agency that apparently the same harm done to a moral patient who is not a moral agent is not really the same harm since it figures differently into the total constellation of characteristics of the beings.

Is there any sound criterion for giving moral consideration for animals? An argument is that this criterion can only be sentience because conscious, sentient beings can be treated rightly or wrongly. The utilitarian, Jeremy Bentham (1748-1832) is often cited: "The question is not, Can they *reason*? nor, Can they *talk*? but, Can they *suffer*?"[33] Our Western ethics emphasizes the significance of pain and suffering. We need take to account of animal suffering. Those beings who can suffer are moral patients, the possessors of inherent value, who are worthy of respectful treatment, of not being treated as mere receptacles. By definition, moral patients, whoever they are, are objects of moral treatments by moral agents. We can be moral or immoral toward them. The group of moral patients is larger than the group of moral agents. Taylor's definition of moral subjects is reasonable. He defines a moral subject as a being that can be treated rightly or wrongly and toward whom moral agents should have duties or and responsibilities.[34] The term "moral subject" may be confusing because "subject" can mean either the focus of attention, as Taylor means, or center of subjective experience, as Regan means with his criterion of the subject-of-a-life. So we may prefer Regan's term "moral patient" to indicate those who can be treated rightly or wrongly, or who are more than mere instruments for us. But the group of moral patients may be larger than the group of sentient animals, larger than the group of subjects-of-a-life. Biocentrists believe that all living beings can be treated rightly or wrongly, and holists or ecocentrists will extend moral concern toward species, ecosystems, Earth, and inanimate parts of nature.

Having postulated inherent value in moral agents and moral patients by using the direct-duty view, Regan wants to find a relevant property which describes all those moral agents and patients who have inherent value. He argues that being a subject-of-a-life is such property. The meaning of the criterion should be understood as serving to mark a distinction between those who have and those who do not have inherent value. Preference utilitarians can also approve the inherent value of those individuals which Regan calls the subjects-of-a-life.[35] We might logically say that if some being lacks all of those mental abilities and hence does not satisfy the subject-of-a-life criterion, then it lacks inherent value. But this is not Regan's argument. For him, the criterion specifies a relevant similarity between moral agents and patients, but it does not differentiate those who have inherent value from those who have not. Not all moral patients, as Regan remarks, satisfy the subject-of-a-life criterion. Those who belong to group (a), that is, those sentient beings which lack other mental abilities, do not. The open question is, do those moral patients have inherent value? Yet Regan hesitates as to whether living but insentient nature can possess inherent value. He aims at rejecting Schweitzer's reverence-for-life attitude in which every living thing has inherent value. Being alive could not be the distinctive criterion between those who have and those who do not have inherent value because we do not have direct duties toward all living things such as potatoes or cancerous tumors.

Regan's theory of inherent value applies to sentient animals. It cannot be applied to non-sentient nature. Inherent value is something that all experiencing subjects have, in accordance with Regan's subject-of-a-life criterion. Many animals can make value-judgments. They are the experiencing subjects of their own lives and have abilities to enjoy their lives. Regan is right in claiming that many mammals have perception, memory, desire, belief, self-consciousness, intention, and a sense of the future. More strongly, all vertebrate animals have the ability to value their own lives. Abilities to express those values in some language should not be required. This is no reason to restrict values merely to sentient animals. The result of Regan's theory of inherent value is that animals do not merit their inherent value by having sophisticated mental states. Those who cannot read or do higher mathematics have equal inherent value. Regan leaves the door open for the inherent value of non-sentient creatures.

7. The Intrinsic Value of Animals

One basic method of revealing nature's intrinsic value is the method of analogy. The analogy is that nature's value has a likeness to the value which people have. That argument, we recall, runs like this: Because human beings have intrinsic value, so do animals since they are sufficiently like humans. The conclusion is that animals have to be treated in relevantly similar ways to people, given that the principle of equality holds. If we postulate the intrinsic value of moral agents, then we cannot deny it to moral patients that are all sentient animals. Animals have to be taken morally into account by analogy with people. Animal rights are seen as the extension of human rights. This extension is supported by arguing that animals are sufficiently like human beings. Ethical extensionism proposes that animal suffering is something to be taken into moral account. Supposing that our responsibility is to promote good in the world, then ethics is extended responsibility to everything that has its own good.

According to the biocentric defense of the good of living beings, human and animal good must be taken into moral account. According to the ideas of zoocentric ethics, forests get their value as being a home for sentient beings. Ethical extensionism finally accords intrinsic value to people, animals, all living beings, and to forests, rivers, and mountains, which are respected and preserved as such. They are not instruments for human purposes but taken into moral account as well as animal suffering and the good of living beings. To destroy individual life forms or ecosystems on the earth is wrong. Conversely, to promote the well-being of individuals and ecosystems is a right deed. The idea of ethical extensionism is that to move from anthropocentrism to zoocentric, biocentric, and ecocentric topics is rational and ethical; to ask what is our ethical relationship with animals and nature is our urgent duty.

Is ethical extensionism an anthropocentric and analogical reasoning from the human cases to the non-human ones on the grounds of non-human similarities with human beings? If so, the center of ethics is the human being and the idea of the intrinsic value of human beings is extended in the first place to animals. Animals are sufficiently like human beings because they have the ability to feel pain and suffering. According to the major Western ethics, suffering must be taken into moral account, if anything must. Hence, animal suffering must be taken into moral account and animal well-being is of intrinsic value which is to be promoted and enhanced.

On the contrary, I do not believe that animal suffering should be taken into moral account for the reason that animals are analogical cases to human beings, or that even plants are relevantly analogous to humans. My thesis is that animals have their own good and well-being, which differ from human good and well-being, and thus must be taken into moral account. My point is not that they are like us but that we and they share a common feature, the ability to suffer and enjoy. My argument is that the intrinsic value of consciousness is a universal intrinsic value which people share with other animals. The intrinsic value of livingness is a second universal intrinsic value which we share with all life forms, and a third universal intrinsic value is nature as a whole of which we are part. Consciousness is an intrinsic value, and *Homo sapiens* is an instance of the group of conscious beings which carry this value. Animals do not acquire their value from the intrinsic value of human consciousness, but human beings get the value of consciousness from the value level of sentient beings. We get our consciousness value from being animals.

According to the invidualistic theory of value, sentient beings as such, without any other attributions, have an intrinsic value. We could in turn argue that no being as such has intrinsic value, but its states, like flourishing or well-being, have this value. Which has intrinsic value: beings themselves or their states? My theory is that we cannot separate beings from their states. The well-being of an animal cannot be ontologically separated from the animal experiences, although we can make this distinction from the conceptual point of view. No flourishing or well-being exists without the holders of these states. Further, an animal's well-being requires the environment. The value of conscious states is the value which is significant in the interactions of animals with their environments. Thus, welfare is a characteristic of an individual in its relation to the environment.

My final suggestion in this section is that the intrinsic value of animals is the value which they possess on the grounds of absolute qualifications from their own point of view. A zoocentric definition is:

> An animal has intrinsic value in itself on the grounds of its inherent qualitative nature from the point of view of its well-being.

In general, this can be formulated as follows:

> X has intrinsic value K on the grounds L from the point of view M.

Label X varies from animals to life forms, and to nature as a whole. Label K can indicate, for instance, experienced-value, self-value, value for its own sake, value in itself. Labels L and M vary; some possibilities for them are discussed in the present book.

From the zoocentric perspective, the well-being of animals has intrinsic value. This intrinsic value is discovered in the well-being of animals, not only in human attitudes. Suffering is intrinsically bad for animals, while well-being is intrinsically good for them. Animal suffering is bad for animals, not only for human beings. People can gain knowledge of animal suffering or well-being from within their value consciousness. Although only conscious beings can get conscious knowledge of values, it does not follow from this that all values are inside conscious beings.

Intrinsic value can exist outside my body and brain, although I get the knowledge of them by looking out from both my body and brain.

Describing animals in terms of intrinsic value is coherent with the aim of preserving and respecting them, while describing them only in terms of instrumental value may lead to the aim of exploiting and consuming them. We should perceive and understand the life of animals in an ethical way. We should answer morally the question of what animals are by their nature. Valuing animals exclusively instrumentally means seeing them as raw-material for us, while valuing animals intrinsically is to confirm their nature as individual sentient beings. Seeing animals as raw-material is unfair from the ethical point of view by giving us a fundamentally false picture of the nature of animals. Animals are not simple raw-material for us but complex sentient beings, according to the zoocentric view that respects animals. If animal welfare is studied mostly from the perspective of instrumental value, our understanding of animal welfare will inevitably be insufficient. New studies should be focused on animal studies in the humanities and social sciences. The many relations of people to animals are largely unknown. Our cultural, political, and religious relationships to animals should be further studied. The journal, *Society and Animals*, is an example of increasing social-scientific interest in animals from the perspective of human relationships with animals. This journal calls for methods and language that respect animals. No conventional animal-experiment methods are reported by the publication, and animals should be called by the terms "she" and "he" to confirm their sentient, individual nature instead of regarding them as tools and raw-material for scientific or commercial purposes.

For animal preservation we need research into the philosophy of what animals are and into the applied animal sciences. Animal preservation has not yet been made as scientific as nature conservation, insofar as different kinds of environmental sciences which aim to protect the environment exist but applied animal sciences are still mostly based on "how to utilize animals" instead of "how to preserve animals." Animal-welfare sciences should take this as one of their aims and seek better solutions for animal-preservation problems. Scientific models and methods of animal preservation should be developed in the social sciences and humanities as well as in applied and natural sciences. The intrinsic-value view of animals is a more sustainable starting point for such research than the instrumental-value view of animals.

In this chapter, I have argued that animals have intrinsic value in terms of their subjective experiences and well-being. In the folloving chapters, I analyze the biocentric and ecocentric position of values.

The basic meanings of intrinsic value	*Animals having intrinsic value*	*Life forms having intrinsic value*	*Ecosystems having intrinsic value*
(1) Experienced or sensed value	1. Animal pleasure, interests, conscious states	No	No
(2) Ends-value X valued for its own sake (anthropogenic value)	2. Can be: pet animals	3. Can be: biodiversity, endangered species	4. Can be: natural beauty, wilderness
(3) Self-value The subject of its own life	5. Yes sentient animals	No/uncertain	No
(4) Inner or inherited value X intrinsically valuable in itself (naturogenic value)	6. Zoogenic value: animal well-being	7. Biogenic value: the good of a living being, survival value	8. Ecogenic value: nature generating value, creative nature

Five

BIOCENTRISM

Many philosophers have restricted intrinsic value to sentient animals. Bolof Stridbeck offers a view of experiences having intrinsic value. [1] Angelika Krebs has defended the view that the scope of human morality should be restricted to sentient beings.[2] Do we have any reason to think that only conscious states or experiential states, such as pleasure and happiness, have intrinsic value? They have such value, but they are not the only valuable states. I prefer a pluralistic value-theory. In addition to conscious states, intrinsic values emerge from many different things, beings, and states. Ethical extensionism is an articulation of the idea of respect for animals, life-forms, and ecosystems. The term "biocentrism" originates in Greek "bios," basically, meaning respect for life. I propose a variety of understandings of life other than the biological model of life. "Life" is a a philosophical concept. It can be used in quite different contexts, for instance, in the following concept pairs:

> the quality of life – Gross National Product (GNP)
> personal life – social life
> material life – cultural life
> eternal life – mortal life.

What we mean by the term "life" depends on the context in which it is used. Consider the following sentences:

> Wildlife should be saved.
> Save the life-support systems of the earth.
> The earth is alive.
> Is country life a more ecological lifestyle than town life?
> Plain living is the best life.
> Her life is no longer worth living.
> What is your attitude toward life?
> What is the meaning of life?
> Abortion is a crime against life.

In these sentences, life is more than a biological fact. The biological definition of life has little sense in these contexts. "Life" is a term that is used and should be understood in its social and ethical contexts. By human life we do not mean the cell, organism, or ecosystem level of life, although from the biological perspective these levels exist in human life. The conscious, social, and cultural level of life cannot be reduced to the biological level. A human life includes the history of the individual, her or his relations to other people, animals, and to the environment. My life is the totality of my personal history and the history of my ancestors which both form my contemporary personality. My life has different aspects; no one aspect can explain my life. Historical, cultural, social, moral, religious, aesthetic, political, biological, physiological, or chemical aspects form the totality of my life. From a more general point of view, this complex understanding of human life is similar to the case of animal life. Animal life has different aspects (historical, cultural, social, biological), and no one aspect alone can explain animal life. Even plant life has different aspects. Biology is the human understanding of plant life, but it is not the only way to understand plant life. Throughout history and in other societies plant life has been considered from a different perspective. All scientific understanding (including the humanities and social sciences) differ from non-scientific understanding such as religious or aesthetic understanding. Another aspect can be revealed from the point of view of nature: what are plants from an animal's point of view? Do plants have any point of view of their own?

Different ideas and concepts of life, including religious, historical, or philosophical notions, provide different aspects of the question, "what is life?" The concept of life is an elusive concept. It refers to no single ontological entity. No ontological entity called "life" exists, but "life" is an adjective, a qualitative property, that can be attributed to several entities. For example, a living being is an ontological entity, but its life includes the history and future of its life. Its life cannot be separated from the individual that it is.

The idea of "save wildlife" may contain a holistic understanding of life in that here is no sense of reducing life to the level of cells or of organisms. We are not concerned so much with plants or animals but with whole wildlife areas. "Wildlife" means the ecosystemic level of life, which cannot be reduced to its subordinate levels or parts. This level contains its greenness, its diversity of life forms, its large areas, its different value-dimensions, including economic, biological, and philosophical values, which make it worth being preserved.

1. What Is Life?

What is meant by such conceptions as "respect for life," "the value of life," and "right to life"? Is all life worthy to be preserved? What we think of life influences our relationships with and responsibilities for other living beings. Do we treat living beings as objects to be manipulated instead of respecting them? Biocentrism, a life-centered ethics, means respect for life, but what is this life that we should respect? How can living organisms be distinguished from non-living entities? Different life forms exist: fauna, flora, fungi, and bacteria. According to the Gaia hypothesis, the whole earth is a living organism with its organic processes.

The following conceptual pairs are synonymous: living-non-living, animate-inanimate, and organic-inorganic. What are the differences between living, animate, and organic, and correspondingly between non-living, inanimate, and inorganic? To what kinds of things can these terms be attributed? Living entities, such as animals, plants, and fungi, are organic entities. Not all organic entities, for example, humus, are living entities. The answer to the question, "What is life?", depends on an attitude toward life. Deep ecologists hold that water, soil, even rocks, are living entities. Typically animals and plants are living and animate entities, while rocks and stones are non-living and inanimate entities. Living and animate are synonymous, just as non-living and inanimate are synonymous. The opposites of living and animate are dead and inert. Living and animate refer to something active, whereas inanimate refers to something inactive. Animate means living and vivified. Living refers to vital. A new hierarchy for the terms "organic," "living," and "animate" would be that organic entities include the largest groups of entities, such as ecosystems and even the whole earth, whereas living beings would include all fauna, flora, and fungi, while animate life would include only animal life. For many people, animate is synonymous with living, including flora and fungi. To understand life we need its opposite, death. From this perspective, different concept pairs are extinct-living, dead-living, dead-alive, dead-vital, death-life, death-birth. These concept pairs as well as animate-inanimate, living-non-living, and organic-inorganic are the key biological concepts of life.

A distinction can be made between "being alive" and "having a life."[3] Having a life requires complex mental abilities. Being alive, the opposite of being dead, is a notion of biology, whereas having a life is a notion of biography. People and animals have their lives in the biographical, historical sense: they have past and future, they have hopes and fears concerning their own lives. Non-human sentient animals are the subjects of biographical lives. In this biographical sense, "having a life" cannot be attributed to plants, although plants are living beings. Plants have biological but not biographical life.

2. The Principle of Plenitude

The world where angels and rats dwell is better than the world where only angels dwell, said Saint Thomas Aquinas. The notion of the diversity of life is one of the most powerful recent ideas of nature preservation. The value of the flourishing of diverse life forms is a basic principle in deep ecology. Biodiversity, the value of life forms, is to be protected and preserved. Biodiversity is a biological fact that can be measured, and it is a value-laden conception. Even if some life forms exist on other planets, our planet remains unique in its varieties of life forms. Should we respect this unique character of our planet? Respect for life is respect for individual lives, and for our planet's life forms. The unique character of life is its ability to vary, to form new, different, life forms.

The best-known contemporary author on the issue of biodiversity is E. O. Wilson.[4] Biodiversity, in its common meaning, refers to different levels of life: to genes, to species, and to ecosystems.[5] Accordingly, biodiversity has three dimensions: the diversity of genes, the diversity of species, and the diversity of ecosystems. The diversity of genes is necessary for reproduction in the struggle of life and death. The diversity of genes produces different individuals, a few of which have more possibilities of survival in different, changeable environmental conditions than if all reproductions were copies of their mother. Most mutations are worthless or neutral to survival, many are detrimental. The diversity of species confirms continuing life on the earth. All species have to disappear as well as all individuals, but new species are developed in the great chain of being. Species are products of evolution. The diversity of ecosystems is necessary to support the variety of species. Different life forms require different living conditions.

A plausible way of examining the concept of life is the old historical and philosophical understanding of life. The idea of diversity culminated in the principle of plenitude in the eighteenth century. Arthur Lovejoy saw that the discovery of the intrinsic value of diversity was one of the greatest achievements of the human mind.[6] The principles of plenitude and continuity are the philosophical diversitarian ideals, although extremely anthropocentric in their history. The principle of plenitude means that the world is the better the more diversity it contains that concerns people (or only men). Plenitude manifests the possibilities of differentness in human nature. The duty of the individual was to cherish and intensify the individual's differentness from other people. This suggested that the first and great commandment was: be yourself, be unique![7] The value of being original was related to the idea of the expansive process of life: the world manifested itself in the maximal differentiation of the beings. Diversity in the multitude of living creatures is the fullness of diversity that is good in itself. The order of the world is not a static diversity, but a process increasing its diversification. Plenitude is a part of good. The best of worlds was the most variegated: all possibilities of life should be realized.[8]

3. Will to Live

Life is more than solid material, it is more than physical matter. Life involves intrinsic power to live, *élan vital*. Arthur Schopenhauer's notion of will-to-live is a forerunner of Henri Bergson's vitalism, the notion of *élan vital*, as well as of modern neovitalism if biocentric views of life can be so called. The source of Albert Schweitzer's ethics of reverence for life ethic is Schopenhauer's idea of will to live. Schopenhauer's will-to-live is not an ethical principle as is Schweitzer's reverence for life. Will-to-live is the essence of world, the thing-in-itself, whereas the individuals are illusions. Life has no purpose for something. Will-to-live is the deepest principle of the world, not explicable by any deeper explanation. In *The Will to Live*, Schopenhauer argues that human life is a mistake. No value appears in it, because it ends in nothing.[9] Boredom is evidence that life has no intrinsic value. Mere existence as such should satisfy us; we should have no interests or wants. This does not apply to the life of animals and plants, because brutes do not suffer boredom in their natural states. Schopenhauer speculates on human life: life is a task to be done. He asks why this world exists rather than nothing. The world does not exist for its own sake or for its own advantage. Life is without ground, a blind will to live. Life is more vanity and suffering than it is worth and pleasure.[10] For Schopenhauer, the world is nothing, life is suffering, and the individual ego is an illusion.[11] Nothingness is the only objective element in time; it is the inner nature of things. The will of entities exhibits itself as sexual impulse, creating an endless series of generations.[12] The will is the *natura naturans*, the inner being of nature. The will-to-live is the thing-in-itself. Schopenhauer calls "will" the force of nature, that which acts and strives in nature. The will is identical with the vital force. For Schopenhauer, individuals are always only means, while the species is the end. Life's intention is the maintenance of the species.

Schopenhauer's idea of the will-to-live is nearer to Bergson's vitalism than to Schweitzer's reverence for life ethics. Thought not primarily an ethical principle, it explains the basic feature of the world. It answers the question of what life is, but not the question of what the value of life is. Life is the will-to-live. The will-to-live is the primary substance, it is the thing-in-itself, the fundamental substance of the world or of the cosmos. Schopenhauer's philosophy is the philosophy of nothingness. The fundamental nature of human beings as well as all other living things is to move from nothingness through life (an illusion) to death, back to nothingness. Many thousands of living beings can die, but nothing of value happens. The world is just the same as it was with those living beings which no longer exist. The will-to-live has no purpose, such as the existence or pleasure of human beings. The will-to-live of human beings is just as blind as it is in animals. It is its own inexplicable purpose. Similarly Bergson's idea of *élan vital* is the final explanation that cannot be explained further. It is the substance of the world.

4. *Élan Vital*

Henri Bergson developed his idea of *élan vital* (life force or internal strivings) in opposition to the philosophical idea of materialism. According to materialism, life and consciousness can be explained in the same terms as the nature of inanimate things. Thus, materialism ignores the special nature of living beings. Bergson differentiates conceptually between the physical, the chemical, and the vital, but he argues that these are not easily separated ontologically. The philosophy of materialistic mechanism is based on the belief that the living can be treated as the inert. Life is no value in this philosophy. Bergson has a pessimistic view of the possibilities of science to understand life in other ways. Science continues to treat life similarly to inert matter. Life is merely instrument.[13] The ethical side, how to treat living beings, is meaningful for Bergson. The flaw in the materialistic and mechanistic philosophy of life is that it gives the wrong picture of the world. Nature does not work like a human engineer. It does not bring parts together, as the development of an embryo shows. Mechanism gives a flawed picture of life. Mechanical causality cannot explain vital, living processes. The sympathetic communication within nature would reveal the true nature of life, which is reciprocal, continued creation.[14] Bergson's vitalism provides a model of life which offers an ethical code that a living being should not be treated as inert matter. His reverential model of life is based on the intrinsic value of life instead of the instrumental conception of life. But while Bergson's vitalism shows that livingness and mentality are not properties of machines, it leaves unexplained how positively to characterize living organisms.[15]

5. The Gaia Hypothesis

A recent form of vitalistic thinking appears in the Gaia hypothesis, developed by J. E. Lovelock, of the origin and nature of life on the earth.[16] The Gaia hypothesis argues that all living things are parts and partners of Gaia, a vast being who has the power to maintain the planet Earth as a fit and comfortable habitat for life. The Earth's atmosphere is actively maintained and regulated by life on the surface, that is, by the biosphere. Lovelock has developed his idea of Gaia and life with Lynn Margulis. They define Gaia as a complex entity, involving the Earth's biosphere, atmosphere, oceans, and soil. Gaia is the totality of this planet that forms a feedback or cybernetic system which seeks optimal physical and chemical conditions for life.[17] Gaia is the largest living creature on Earth, which consists of the life-supporting level of planet Earth. An almost infinite variety of living forms proliferates over the Earth's surface. Our planet has unique life-support systems, and without them no life would exist. It is not a dead planet but a living planet. The planet Earth is different from the other planets in its ability to maintain life. The surface, the biosphere, of the planet Earth constitutes a living organism, Gaia. The surface of our planet is a vast ecosystem that can be called the largest living entity. This is the sense in which Arne Næss and other deep ecologists understand life. The Gaia hypothesis is not the widest sense of the term "life." In the widest sense all material existence, the whole cosmos, is a living entity.[18] Our planet is a living planet, the only one in the whole cosmos. But, what we mean by the term "living organism," may be confusing. If life is restricted to fauna, flora, and fungi, then speaking of the Earth as a living being is difficult to understand. For Lovelock, life means more than fauna, flora, and fungi, which contain the physical and chemical systems of the earth. Gaia refers to what we may more scientifically call the "biosphere," the surface of the earth that maintains life, through life-support systems such as chemical and physical cycles (CO_2, O_2), and photosynthesis.

All our conceptions of life have their limits. The mechanistic conception of nature is no less a vision than the reverential conception of nature.[19] The Gaia hypothesis is a reverential conception of life. The value of the Gaia hypothesis is that it raises the question of life as a fundamental problem of our modern culture. Whatever we make of the Gaia hypothesis, no one can deny that Earth is a prolific planet, nor that this phenomenon of life is of fundamental value. Similar to vitalism, the Gaia hypothesis is right in criticizing the mechanistic view of life. Life is the novel characteristic of our planet that differentiates it from others. But while saying that the earth is alive is true, Lovelock's Gaia hypothesis does not explain the origins and characteristics of life.

6. The Ecological Model of Life

In ecology, the cell level is called molecular ecology, the organism level is organic ecology, and the population level is population ecology. How do living beings differ from non-living ones? The modern biological understanding of life is in terms of levels: the cell, the organism, and the population. The beginning of life is in two molecules: deoxyribonucleic acid (DNA) and protein. The DNA molecules of the cell form the internal designing forces of life.[20] The basic unit of life, regarded as the origin of life, is the cell, because the cell reproduces and metabolizes as an independent unit. Biologists have noticed that cells contain small entities, organelles, such as plastids and mitochondria, which have been described as subvital. In 1935 W. M. Stanley isolated a virus in crystalline form.[21] Living beings can reproduce themselves. The living organism is made of macromolecules: carbohydrates, fats, nucleic acids, and proteins. No machines are made of these. The organism owes its structure to internal designing forces. The developmental process from zygote to horse is internal to the organism.[22] We can therefore distinguish an animal from a machine. The machine cannot reproduce itself, whereas the animal can. The animal is made of those macromolecules (carbohydrates, fats, proteins, nucleic acids, etc.), but the machine is made of small molecules consisting of components of iron, nickel, and aluminium. The machine could be made of macromolecules, such as polythene plastics, but no machine is made of carbohydrates, fats, proteins, and nucleic acids. The living animal has a complex metabolism and internal designing forces, while the machine owes its structure to external forces, such as its designer and manufacturer.[23]

The ecological model of life stands against mechanistic explanations of life.[24] Why could not mechanistic explanations tell the truth of life? The answer is that Newtonian science was dealt with external relations. The model of a machine explained how material substances acted upon one another. The ecological understanding of life involves the relationships between the biological levels of life and the individual's relationship to its environment. The individual organism is a product of the internal forces of its genes and of the external forces of the environment.[25]

7. Reverence for Life by Schweitzer, Næss, and Skolimowski

The ecological model of life calls attention to the reverential relationships between people and nature, like Bergson's vitalism and the Gaia hypothesis. The principle of reverence for life does not imply Schopenhauer's idea of the will-to-live. Schopenhauer denies any intrinsic value in life. Thus, he gives no basis for conservation ethics. I believe that the notion of "will-to-live" is fundamental in ethics. The modern *ethical* defense of will-to-live is Albert Schweitzer's well-known ethics of reverence for life, whose principle is: "I am life which wills to live, and I exist in the midst of life which wills to live."[26] This is the ethical conception of life. To manifest life is to manifest the will-to-live. The fundamental principle of morality is: "It is *good* to maintain and cherish life; it is *evil* to destroy and to check life."[27] The main ideas of Schweitzer's ethics of reverence for life are: (1) life is sacred; (2) all living beings belong together; (3) the reverence for life attitude includes all real values, such as sympathy and love. This is ethical extensionism, ethics being extended responsibility to everything that has life. If I find my life an intrinsic value, then I understand that other lives will to live and have intrinsic value.

Arne Næss is the founder of deep ecology in his early philosophical defense for life.[28] Næss's ecological principle is: live and let live. The principles of deep ecology are formulated by Næss and George Sessions together. The first and the deepest principle of deep ecology is the well-being and flourishing of both human and non-human life. This locates intrinsic value in the well-being and flourishing of life-forms on Earth. Life-forms have value in themselves independent of their usefulness for human purposes.[29] Developed humanity leads to identification with animals and nature, all life forms having intrinsic value in the wide sense of life.[30] The term "ecosphere" could be used instead of the term "biosphere," to stress that "life" refers to such things as rivers, landscapes, cultures, ecosystems, and the living earth.[31] In the narrow sense of life, these entities would be classified as non-living. The broader usage of the term "life" appears in such slogans as "let the river live." This is not just a metaphorical expression. A river can be perceived as a living entity, this "living" having a broader value-meaning than in the biological sense.

The value of life is the fundamental principle in deep ecology. The well-being and flourishing of human and non-human life have intrinsic value independent of the value that life-forms have for human purposes. Richness and diversity of life-forms are values in themselves and contribute to the flourishing of human and non-human life on Earth.[32] Two ultimate norms of deep ecology are self-realization and biocentric equality: "all things in the biosphere have an equal right to live and blossom and to reach their own individual forms of unfolding and self-realization within the larger Self-realization."[33] The idea of self-realization is a fundamental norm of deep ecological thinking. The basic idea is that all living individuals have their own self-realization, so that all individuals are parts of Nature's Self-realization. Deep ecology describes and rejects the Western view of "self" as a hedonistic, lonely, and isolated ego."[34] The manifestations of life should be maximized. A principle of the self-realization involves an intrinsic value in diversity. Natural diversity and plant species have their intrinsic value, while shallow ecology sees in them valuable resources as genetic reserves for human agriculture and medicine. In deep ecology, talking about value as value solely for people is nonsense, because for deep ecologists, "resource" means resource for living beings. Both intrinsic and instrumental values exist in the non-human world.[35]

Henryk Skolimowski, the author of *Living Philosophy*, uses the terms "eco-ethics" and "ecological ethics" as synonymous.[36] His philosophy is called eco-philosophy, in accordance with his earlier work.[37] The question of life is fundamental for his ethics. He has even said that environmental ethics is an extension and articulation of the idea of reverence for life.[38] He differentiates first-order, second-order, and third-order values. The first-order values are foundation values. The second-order values are the consequences of foundation values. Specific tactics and strategies to achieve these values are called third-order values. By asking "why?", ethics leads to foundation values, to these ultimate norms that guide our actions. Skolimowski's new moral insight was the recognition that nature should not be an object trampled upon.[39] Animals and plants were not created for human purposes and uses, although we are a part of a living nature. For Skolimowski, the basic values of eco-ethics are reverence for life, responsibility for our own lives, frugality as a precondition of inner beauty, the pursuit of wisdom, self-actualization, and understanding of the heritage of life and of evolution.[40] A resulting ethical imperative is to behave in a way that preserves and enhances ecosystems as a necessary condition for life and consciousness.[41]

8. Paul W. Taylor's Theory of Inherent Worth

Paul W. Taylor compares the anthropocentric attitude to the biocentric attitude: biocentrism involves respect for living nature in the same way as anthropocentrism involves respect for humans. The concept of inherent worth is a part of Taylor's biocentrism. Inherent worth is the foundation of the moral considerability of each living organism. This is based on each living being having a good of its own, which is the objective perspective of inherent worth. The good of a living being is an empirical fact of this being. Knowledge of the good of living beings comes from biological and other studies; thus, we are able to take the standpoint of each living being. If we consider the standpoint of each living being as morally significant, then we have made a value-judgment; this is the subjective perspective of inherent worth. Taylor distinguishes inherent worth from inherent value and intrinsic value. Inherent value is likened to ends-values or it is seen in opposition to instrumental values. Intrinsic value is linked to states of experiences, in accordance with the theory of value of C. I. Lewis. I begin with the Lewis's theory of value, which is approved by Taylor, and then analyze Taylor's concepts of different kinds of "i-values": intrinsic value, inherent value, and inherent worth. At the end of this chapter, I compare inherent worth to Regan's concept of inherent value.

A. Intrinsic Value in Terms of Experienced States, Inherent Value as Located to the Objects of Experienced States, and Inherent Worth

The sub-classification of extrinsic values originates in C. I. Lewis's *An Analysis of Knowledge and Valuation*.[42] Taylor has formulated Lewis's distinction in the following way:[43]

(1) intrinsic value [=experienced-value]
(2) extrinsic value
- (a) inherent [=ends-value]
- (b) instrumental
- (c) contributive

According to Lewis's theory of value, intrinsic value is necessarily connected to states of experience. In the following, I symbolize this intrinsic value by "intrinsic $value_L$" which means "intrinsic value in terms of experienced states according to Lewis" to distinguish this intrinsic $value_L$ from the varieties of meanings of "intrinsic value" in this book. Lewis's sense of intrinsic value is not a useful candidate for what is the intrinsic value of nature, because this form of value is restricted, if not to human beings, then to beings having experiences. Inherent value in Lewis's sense can be attributed to things which are valued for their own characteristics, like the objects of arts, or a beautiful landscape. I call this kind of value "ends-value." I symbolize it "inherent $value_L$" which means "inherent value according to Lewis."

Intrinsic $value_L$ is a non-derivative value. This has been expressed well by C. A. Baylis:

> For any entity to be *intrinsically good* it must have a non-derivative positive worth; it must be good in itself; it must have positive value over and above any extrinsic value it may have. It must not owe all its value to its relations to other valuable things. If an experience, or anything else, has intrinsic value this value belongs to it in virtue of its own nature and is independent of any derivative value it may possess.[44]

Extrinsic values derive their value from something which has intrinsic value$_L$. They may contribute to the existence of something which is inherently valuable$_L$; but an inherently valuable$_L$ thing actualizes its value only when it is experienced. In this view, objects, acts, situations, or persons can have only extrinsic value, because only a felt or perceived quality of experience have intrinsic value.[45]

Taylor follows Lewis and Baylis as regards this classification of intrinsic and extrinsic value. In addition to,these, he has postulated a new type of value: inherent worth. Taylor has developed his theory of inherent worth in *Respect for Nature: A Theory of Environmental Ethics*, "The Ethics of Respect for Nature," "In Defense of Biocentrism," and "Are Humans Superior to Animals and Plants?".[46] Taylor's major work on moral philosophy is *Normative Discourse*, and he is the editor of *Problems of Moral Philosophy*.[47]

In "Are Humans Superior to Animals and Plants," Taylor has not yet made the distinction between intrinsic and inherent value which he makes in *Respect for Nature*.[48] In 1984, he classifies works of arts, areas of wilderness, or historically significant objects into the group of intrinsically-valued entities. In *Respect for Nature*, inherent worth is distinguished from intrinsic value and from inherent value. Intrinsic value is related to conscious states as the classification of intrinsic value by Lewis claims. Enjoyable experiences have intrinsic value. Experience could have extrinsic value. But when it is valued for its own characteristics, then it is valued as intrinsically good and has intrinsic value$_L$. Inherent value$_L$ differs from intrinsic value$_L$ with respect to the object which has this value. Intrinsic value$_L$ is located in conscious states; inherent value$_L$ is located in objects outside conscious states. Taylor lists those things we want to preserve as such, independent of their possible derivative value, which therefore have inherent value$_L$: works of art, historical buildings or places, battlefields, wonders of nature. Living things, like pets, can belong to this class of value. The inherent value$_L$ of an object is dependent upon valuation.[49] The value of a beautiful landscape is inherent in that I (or other conscious beings) value it for its beauty. The value appears in the relation between myself and the object which I like or choose for its own characteristics. Intrinsic value$_L$ is "intrinsic" (inner, inside), because it is located in the states of mind, in experiences. Inherent value$_L$ is extrinsic because of its derivative nature. A beautiful place of nature derives its inherent value$_L$ from the enjoyable experience it gives us. Taylor defines inherent worth as follows. If an entity has a good of its own, and if it is better that this good be realized than not, then the entity has inherent worth. The good of living beings and their inherent worth is the fundamental value for the attitude of respect for nature.[50] Taylor holds that the concept of inherent worth is essentially identical with Regan's concept of inherent value. Like Regan, he separates inherent worth from values of any other beings and from merits. Those who have inherent worth have it in their own nature, and it belongs to them inherently. Animals and plants deserve our moral concern and consideration and should be respected because they have inherent worth.[51]

A basic feature is the absence of a reference to the valuation of valuing subject. What does Taylor mean by saying that "a state of affairs in which the good of X is realized is better than an otherwise similar state of affairs in which it is not realized" in his definition of inherent worth?[52] This betterness is binding on us if we have accepted this definition of inherent worth. In whose minds will it be better that the good of entities is realized? If this is a human value judgment, does the inherent worth of a pig disappear, when somebody decides to eat it? For the pig, not to be eaten is better. Taylor's idea must be that this betterness refers to the same being whose own good is at stake. For a living entity, it is better that its good be realized than not realized.

B. Criticism of Taylor

The basic problem raised in Taylor's concept of inherent worth, is the difficulty of showing that all living beings have a good of their own, while non-living entities do not have a good of their own. Why should being-alive be a criterion which implies the concept of "the good of a being" for whatever lives? Taylor does not give fully sound arguments for this demand. A car has no such good of its own as being well-oiled; the driver's interest is to keep the car in good condition. The same may hold true of plants. Plants have no good of their own in the same sense as I have a good of my own, although the gardener's interest is that plants flourish.

Some philosophers have believed that we have no reasonable grounds to argue that plants have a good of their own. Values are related to valuers. No good can exist without a subject for whom something is good. What could be the good of a garden, if it is not the good of anybody? If everything can be good, then "good" qualifies nothing.[53] If X does not care for how it is treated, then we cannot treat X morally wrong. A plant does not worry about itself no more than does a car, so it cannot be treated wrongly. Values can be centered on the non-human nature, but these values do not exist independent of conscious valuers.[54]

Many biologists and nature conservationists may find it conceptually, ontologically, and morally hard to believe that individual plants have their own good that we should respect. The analogy between respect for individual human beings and respect for individual plants does not function in practice. From the practical point of view, it may be impossible to take account of the good of every living plant and animal. In Taylor's view, even grass has inherent worth and deserves moral consideration equal to the good of a human individual.[55]

A plant's good would be understood in terms of the flourishing of living beings. The problem is that although the flourishing of living beings depends on their own nature, it does not follow that their flourishing has intrinsic value. We may say that the flourishing of beings is one of their characteristics, but whether it is valuable is another matter.[56] If we agree that flourishing has value, then this is an instrumental value for living beings. By flourishing, living beings fertilize and produce new generations. We may argue that flourishing is just a means for genes to reproduce. For people, the flourishing of living beings has economic value, like the products of agriculture, which is an instrumental value. Seeds grow or do not grow. We may believe that nothing is good or bad for the seeds themselves whatever happens. Or suppose you have a spider plant, *Chlorophytum*, which produces runners. Does each runner have its own good? Do you damage the runners, if you throw them away? Or suppose you have an apple tree. Is the good of each seed to grow into a tree? Each seed is a teleological center, as Taylor stresses, whose telos is to grow into plants and trees. But all species overproduce, including people.[57] The life-and-death roles are in the balance of nature. Many lower and higher organisms die prematurely from diseases, starvation, or predation by other organisms.

For us to understand what the good of an organism is requires that we be something like that organism. Nagel's challenge of what it is like to be an organism, is analogous to the question of what the good for or of the organism is. Nagel does not speak of the good of a being , because he has no moral view in his discussion.) In order that something could be good or bad for the organism, there must be how it is for the subject itself. This is the subjective character of experience.[58] If something were "my own," "I" would exist which experiences something as its own, and which has the ability to separate itself from others. If I have a good of my own, I should be able to know what the good for me is, which presupposes that I can differentiate the good from the bad. If I do not know this difference, I cannot know whether something is good for me or not. From being alive it does not follow that the entity-being-alive is able to make distinctions between good or bad, welfare or illness, self or others. The plant does not care or know whether it is watered, because the plant has no self, mind, and no experience of itself. Because the plant does not have ability to make the distinction between being watered and being unwatered. Therefore, plants have no good of their "own" in the same sense than I have a good of my own.

Taylor's individual-centered biocentrism runs counter to the holistic arguments of species preservation, or the Leopoldian view according to which the ecosystems and populations are more worthy and significant than individual animals or plants.[59] Taylor could reply that ecosystems need to be preserved just to preserve individuals and facilitate the lives of future ones. The problem concerns the basic issue of preservation: do we preserve individuals, species, or ecosystems?

Yet non-living entities as species or ecosystems can be said to have their own good in reference to their well-being. From the view of preservation of species, in our ordinary environmental language we can damage or protect individuals and ecosystems. A real concern arises about how to protect life-support mechanisms like the ozone layer from the damage caused by human pollution. Another example is that environmentalists and biologists may not show much concern for the damage to individual trees which foresters cut down, but they may be much concerned about multistress diseases and other causes that kill whole forest communities. We can speak of the welfare of ecosystems or communities. This is the point made by von Wright in *The Varieties of Goodness* and by Taylor in *Respect for Nature*, but in their view the good of communities will be reducible to, or realized in, the good of their individual members.[60] In Taylor's view, a good of its own is applied only to individuals, not to populations or communities: "The population has no good of its own, independently of the good of its members."[61] But we can speculate whether ecosystems have a non-derivative welfare, too.

If the arguments above are right, then being alive as plants cannot be the distinctive criterion for entities to possess their "own" good. Mechanics know what is harmful or beneficial to the motor of a car, and wood technicians know what kind of conditions, such as temperature and humidity, are good or bad for different kinds of wooden material. Being alive or not does not make the difference here. From the zoocentric point of view, we may stress that the good of animals and the good of plants are completely two morally different cases because plants do not resemble animals in their ability to suffer or feel pain. Animals have a good of their own, because they are experiencing and suffering subjects of their own lives. We may differentiate psychic, physical, and social features of the good of an individual animal. The welfare of animals is an accepted topic of science, and no normally educated human being denies that animals can suffer or experience pleasure (on animal suffering and well-being see Chapter Four).

C. The Problem of the Method of Analogy

The basic method of revealing nature's possible value is the method of analogy. If we postulate the intrinsic value of moral agents, then we cannot deny it to moral subjects. To defend nature's intrinsic value beginning from a human being's value is difficult, because human beings are only a part of nature, and a considerably atypical part of nature. From the supposed intrinsic value of human beings it does not follow that other parts of nature or the whole of nature is of intrinsic value. This would be an inductive type of reasoning in which the conclusion is not deduced from the premise. The passage from the intrinsic value of human beings to nature's intrinsic value is problematic. If we can confirm that people are in some distinctive way intrinsically valuable, then we have no logical way to get from this is to the intrinsic value of nature. This is the indirect method of revealing nature's intrinsic value based on the analogy of intrinsic value of human beings to nature's intrinsic value. Conversely, we should show that all living beings have intrinsic value. Then people would have intrinsic value as a part of the living nature.

As far as moral standing is based on a human paradigm having interests and well-being, then no logical necessity exists that the basis of moral standing be the same in the case of different species of beings. If moral standing belongs to each human being, it does not follow that all other beings have moral standing because they have analogical characteristics to people. From the case of people, we move the argument to something which concerns all living beings. A moral presupposition is that similar cases should be considered in the same way. What would be the sufficient similarity between all living beings? It may be suggested that having interests and one's own good is a substantial similarity between different kinds of living beings, although at first glance it looks like a dissimilarity between us and those beings which have abilities different from ours. An example is marginal people. Severely malformed or retarded people still have moral standing. The analogy here is not that non-human beings count in morality because of their similarity with abnormal people, but because of the similarity which all those creatures share. Normal people, abnormal people, animals, and all living creatures have their own interests. They can flourish or perish; they have a good of their own. These properties give them moral standing.

Although interests and one's good would be common to each living being (which may not be true of many plants and animals), we may yet doubt whether they should have moral standing. You can argue that human beings have moral standing because they belong to human communities, not because they share some substantial characteristics. You can even argue that marginal people are still human beings because they are members within human species and one of ours in a sense that animals are not.[62] We may doubt whether all living beings have interests or their own good analogously to people. G. H. Paske uses the concept "being an end-in-one's-self" which he connects to the moral standing of beings: subjects being ends-in-themselves have rights.[63] While some thinkers observe similarities between people and all living beings, Paske observes differences: The difference between people and animals is that a human being is always an end-in-one's-self.[64] The morally relevant difference between human being and other creature is that, at most, only a few animals have conceptually-based emotions and are capable of abstract thought.[65]

The skepticism against even animals' interests appears in M. A. Fox, who does not deny that each animal or plant life is a unique life, though he sees no moral analogy between people and animals.[66] Even if moral standing can be in some way induced or derived from a human characteristics, such as having interests or one's own good, the analogy between human beings and non-human living beings would be insufficient to derive animal's moral standing.

D. A Defense Again

We can seek further arguments for the thesis of the good of living beings. In von Wright's view, the good of a being refers to the being, "who can meaningfully be said to be well or ill, to thrive, to flourish, be happy or miserable."[67] This biological view of the concept of good fits well with Taylor's theory in that all animals and plants have their good. We need to distinguish between the subject-of-its-own-life and something-being-good-in-its-life. Plants are no subjects-of-their-own-life as animals are, but plants can still be good-in-themselves. We can get objective knowledge of the life cycle of a butterfly, an apple tree, or a protozoon, and in this sense we know something of what is good for them. The tendency to employ the expression "good for plants" suggests a biological context. The unity of an organism is a mode of conserving and promoting life.[68] We may assume, following Taylor, that every organism defends its own kind as a good kind. The tree defends a good of its species; thus, it is a valuation system in itself. Its physical and chemical environment must be good for it. In this biological sense, a plant values its environment.

The concept of good is a link between facts and values in Taylor's theory. The good of a being is a biological fact of its being. What is good for the being depends on the being's own nature. Value-concepts are based on such concepts as wants, needs, welfare and illness, happiness and sadness, properties which are psychological, but which are in some sense biological, too.[69] All living beings have needs. A plant needs water because without water it will die. If it is given all the things that it needs, such as water, light, warmth, and nutrients, then the plant will flourish. Without them, it will perish. The concept of the good of plants can be defended fruitfully in terms of being a good of a kind: each entity having its essential goodness, being good as what it is. This is the biological goodness of living beings. I do not mean by biological goodness that plants have something analogous to the good of human beings but that plants have a different good – the good of plants. They are good as such, not because they resemble human beings. We need not think that plants have interests or a good of their own in the same moral sense than I have interests or a good of my own. They have the types of livingness that we should respect, although they are different from us. In their differentnesses, they can increase the varieties of goodness in the world.

9. Comparison Between Regan and Taylor

I will conclude the discussion of the concept of Taylor's inherent worth by comparing inherent worth to Regan's concept of inherent value. These authors have been influential in a fruitful way. Several similarities occur between their individualistic value-theories, although significant differences also appear. Regan's inherent value and Taylor's inherent worth resemble each other on ontological grounds. They agree that some non-human beings (for Regan animals, and for Taylor plants) have a good independently of people, that those beings can be harmed, and that harming them is doing wrong, absent sufficient justification. Regan's subject-of-a-life criterion and Taylor's good-of-beings criterion are good standards for the value of living beings. Taylor's concept of moral subjects corresponds to Regan's concept of moral patients.

Inherent value = zoogenic value (Regan)	*Inherent worth = biogenic value (Taylor)*
moral rights view	taking one's standpoint
the subject-of-a-life-criterion	the good-of-a-living-being-criterion
moral patients: human beings and other sentient animals	moral subjects: all living organisms, wild organisms especially

Both concepts refer to those who deserve our moral concern and consideration. For Regan, the deserving parties are sentient and conscious animals, while for Taylor they are living beings. Regan and Taylor concentrate on the welfare of individuals, not of ecosystems or populations. To take account morally of those moral patients or subjects, Regan constructs the direct-duty-view, while Taylor constructs the respect-for-nature view. They both accept the equality principle that whatever has inherent value or inherent worth has it equally.

Taylor stresses the objective nature of the concept of the good of a being. By objective, Taylor means that all animals and all plants have a good of their own. The good of a being is an empirical fact concerning that being. Subjective value appears when conscious beings appreciate something desirable or undesirable.[70] An example is that an alcoholic wants to drink, but drinking is not objectively good for her or him. Regan separates preference interests and welfare interests that correspond to Taylor's view of subjective and objective values.[71]

Undoubtedly, Regan and Taylor have developed conceptual apparatuses which are successful for understanding the value of animals and nature. They have stipulated inherent value and inherent worth as their own terms. I would say that Regan defends zoogenic value, while Taylor defends what I call biogenic value.

These concepts are more informative, since they indicate the value inhering in animals or in living beings. The value independent of human consciousness is naturogenic value. The voice of Regan and Taylor is close to what I call naturogenic value, value generated by nature. I defend objective value, which is independent of moral agents. Sometimes Regan and Taylor discuss value as an attribution of human beings. Do they intend that the value of animals or living beings is a human attribution to them, the value, which is ascribed or given to them? I do not believe that this is their meaning.

Six

ECOCENTRISM

How wide can ethical extensionism be? Can it be extended from ecosystems to the earth, and finally to nature as a whole? Perhaps we cannot take into account the level of nature as a whole in its widest sense: other planets, space, universe. We do need stop ethical extensionism at the level of ecosystems within the earth. Then, ecocentrism is a final level of ethical extensionism, which is a pluralistic view of the value of nature. Environmental ethics requires an attitude of impersonality: I speak and act for others. In anthropocentric ethics, "for others" refers to people. In naturocentric ethics, the others include animals, living organisms, species, ecosystems, and the whole Earth.

A chapter on ecocentrism cannot begin otherwise than by citing Aldo Leopold, in *A Sand County Almanac*: "A thing is right when it tends to preserve the integrity, stability, and beauty of the biotic community. It is wrong when it tends otherwise."[1] The notion of the natural community as a whole is a key concept of ecocentrism. It extends the idea of community to include living individuals as animals and plants, the land, water, and air, as well as forests, rivers, and mountains.

In this chapter, I introduce some basic ideas on ecocentrism. Ecocentrism means a philosophy in which the issues, concepts, and values of ecosystems are central. I begin with some remarks of anthropocentrism as opposed to the organic world-view. Second is the topic of individualism versus holism. With this philosophical background, I turn to three environmental philosophical topics on ecocentrism: biocentric holism, anthropogenic ecocentrism, and Leopold's land ethics. Finally, I discuss ethical extensionism – ethics is not complete until extended to nature as a whole. We need an ethics for animals and plants, which is one level in a comprehensive environmental ethics. We need to respect both individuals and ecosystems.

1. Instrumentalism Versus the Organic World-View

The instrumentalized world-view is the ideology of strong anthropocentrism which ignores our dependency on nature by mastering and dominating it.[2] The instrumental view of nature has been world-wide. It is taught in most Western schools as our common-sense reality. We have lost the old organic world-view. The revolution of science meant the death of nature. Nature came to seen as a dead, inert, manipulable material. The instrumental view of nature has been successful, since viewing the world as a resource for people is the practical and useful way to organize information and the whole world for the sake of production. In a new urban individualistic lifestyle, artificial elements are created by and for people. Little attention is paid to the life of ecosystems outside of urban areas. Thus, it can be said that modern anthropocentrism neglects nature and environmental effects. The idea of instrumentalism revolutionized the self-conception of natural scientists from servants of nature into manipulators of nature by forcing and controlling nature.[3]

An example of the instrumental-property view is that if you have promised your neighbor to water her or his flowers, your duty concerns only your neighbor, as if you were watering her or him, not the flowers.[4] Meyer-Abich argues for the impossibility of instrumentalism. He is optimistic about the possibility of change. A theory of autonomy justified the view that the world outside be handled for the greatest personal advantage. An idea of Enlightenment was regard for general well-being, that each human being is equally worthy of moral concern.[5] But the Enlightenment never concerned being other than people. Enlightenment was human equality, the idea that we are born equal. The second Enlightenment should be the revolution for nature. It should extend the principle of equality to the non-human world.[6]

Technology or machines may be the problem, in addition to the spirit and soul of mechanistic, instrumental thinking. The instrumentalist view of the world restricts our ability to see nature in other ways. Nature is revealed to us as an object for technological and commercial processing. In the instrumental world-view, animals and nature are treated as tools and raw-materials, not as living beings.

2. Individualism Versus Holism

The following concepts are related to each other: unity, wholeness, integrity, order, diversity, variety, harmony. These are sometimes regarded as the components of beautiful objects.[7] R. B. Perry describes the value of unity in terms of human intelligence: "It may be said that thought seeks to be as all-comprehensive and unified as possible, and that an orderly whole is the goal of its endeavor ... . It proves that unity is *a* value."[8] According to Perry, the value of a physical organism is the reciprocal fitness of the parts which compose a unity having a distinctive quality. The value of the unity of a physical organism lies in the fact that its components function interdependently and in a manner subordinate to the whole, just as the value of beauty in its aesthetic unity is a balanced composition in which the beauty of the whole derives from the interrelation of the parts. Components are understood in terms of one another and in terms of the created whole of which they are parts. This creates the value of wholeness which is not reducible to the value of individuals.[9]

An environmental ethics should take both individuals and wholes into moral account.[10] We should recommend moral pluralism by taking diverse moral principles and responsibilities into account.[11] Trying to judge our relationship with ecosystems on solely individualistic moral grounds cannot suffice. The issue of animal rights is conventionally argued with individualistic arguments, in which individuals have intrinsic value. The animal-rights view offers an argument for nature preservation. Forests and other natural areas are home for many animals, and since interests of animals should have moral standing, their interests to live in peace in their environments should be taken into account.[12] However, this kind of argument does not satisfy those philosophers, who argue for the independent value of natural areas, like wildernesses or mountains, apart their possible values for individual sentient beings. We miss the conservation value of species and ecosystems, if we value only sentient life or only individual life-forms. According to the biodiversity principle, we have no right to destroy ecosystems on the earth. Conversely, we do right in promoting the diversity of species. A complete nature-conservation philosophy should take both individuals and ecosystems into the scope of our morality.

3. Anthropogenic Holism, Biocentric Holism, and Physiocentrism

Three ecocentric positions are J. Baird Callicott's anthropogenic holism, Lawrence Johnson's biocentric holism, and Klaus Meyer-Abich's physiocentrism. Callicott's theory of value should be classified as anthropogenic holism. It is not necessarily anthropocentric. People generate value but do not center it on themselves. Nature as a whole has intrinsic value. All values are necessarily related to conscious states. We can value natural objects for their own sake, but they are valueless without valuers. This argues for the need to distinguish between valuing and the locus of value. Nature, not a valuer, can be the locus of value. Callicott has called this view of value the idea of "truncated intrinsic value."[13] Johnson has recently defended the view that species, ecosystems, and the biosphere as a whole have intrinsic value. This theory is biocentric holism. In his view, species, ecosystems, and the biosphere have interests, and thus they belong to the scope of morality as morally significant entities and have intrinsic value. Johnson extends the moral arena to living entities, but not to non-living entities. For him, only living entities can have interests which can be benefitted or harmed. These living entities are individuals, species, ecosystems, and the biosphere. His example is a hive of bees that has organic unity, self-identity, and the ability to flourish or suffer.[14] He does not make the distinction between animal welfare and the concerns of nature-conservation activists. Johnson's holism should be classified under biocentrism, because according to him only living entities have moral standing in which the distinctive criterion is "being alive" or not. Life is a central concept in his theory.

Meyer-Abich has introduced the word "connatural world," Mitwelt.[15] We and other species live together within the world. Nothing belongs to us, for we belong to the world. People should not be the measure of all things, which have their own values.[16] Meyer-Abich has also developed the notion of physiocentrism, which includes the idea of nature as the connatural world.[17] By physiocentrism I mean, following Meyer-Abich, the view of nature as a *physis* or as a whole. The term "physis" or "physio" originates in Greek and means nature (as a whole), not only physical nature. Yet the scope of physiocentrism is not an easy task to define. What is nature as a whole, and what is its value? In recent times, ethics for nature as a whole has been a controversial area. Many environmental philosophers extend duties to animals and plants. They hesitate to extend ethics to non-living entities such as rocks or ecosystems. Some may argue that the environment as a whole is too big to be taken into moral account.

Could it be that rocks, mountains, and rivers have intrinsic value? Do they have their own good? Can we benefit or harm a mountain or a river? Is it rational and moral to say that they have well-being? Speaking of the well-being of the earth is not nonsense in every case. We can damage ecological and geological systems on earth. We can destroy immense parts of the earth with nuclear bombs, although the physics of the earth will remain. In this sense, the earth can be said to have the state of well-being. Is it completely a different notion of well-being in comparison to the well-being of animals? We can find some support for the idea of the value of nature as a whole in Leopold.

4. Aldo Leopold's Land Ethics

Leopold's key concept of nature is the land. In addition to soil, it refers to whole biotic communities, to ecosystems. Leopold suggests the land as a living being. Instead being a commodity belonging to us, land should be regarded as a loved and respected community to which we belong.[18] The land is a valuable source of energy for plants and animals.[19] The energy of food chains returns to the soil by the death and decay of beings. The land is a basic unit to be respected and preserved as such, in addition its economic utility for humanity. Leopold holds that the extension of ethics should enlarge the concept of community to include soil, water, plants and animals, and finally, the land as a whole.[20] This is a change of attitude from regarding the land as resources to affirming the right of animals, of plants, and of ecosystems to exist in a natural state. Land ethics changes our role from conquering the biocommunity to respecting it.[21] Things regarded as unnatural, tame, and confined should be reappraised in terms of things natural, wild, and free.[22] Leopold defends the concept of land as a community. We should think like a mountain in an objective and ecological sense, not in the sense of personification.[23] The first ethics concerned the relation between individuals, the second ethics dealt with the relation between the individual and society, and the third ethics is the evolution of the land ethic.[24] People are members of a community of interdependent parts.[25]

Leopold came to care about nature, and especially wolves, after he shot an old female and watched a green fire dying in her eyes.[26] For me as a defender of zoocentrism and ethical extensionism, this example as Leopold's attitude to wolves is as macabre as would be to shoot babies in coming to care about them. I need not shoot a human being to understand the intrinsic value of a human life. From the naturocentric perspective, human life and well-being in the world is not the only significant life. In the wild, life of an animal is the significant life for the animal itself. Thus, from a zoocentric position, the wrongness of hunting is not dependent on what motivation hunters may have but on the harm done to these animals, from the animals' point of view.

According to hunters, hunting builds one's character, increases bonds with nature, and provides a positive attitude to nature conservation.[27] This is the anthropocentric attitude that locates values in human attitudes. Human interests are taken into moral account, and no heed is taken of the animal which is shot. We are heedless of the suffering of the injured wolf or the starving cubs and the mate of the dead wolf. The criterion is the intent in the mind of the shooter. With the right intent, the deed is right.

Although Leopold is made a historical symbol as defender of ethical extensionism, his attitude is not the truly zoocentric or naturocentric position. From these positions, harm to animals or to nature matters, not just harm to people. The wrongness of an action does not depend on our motivation or costs and benefits for us, but on costs and benefits for animals and nature. The good of animals and the good of nature is the question. I hope that the above interpretation of Leopold's ethics is not correct. Since Leopold wants us to think like a mountain, perhaps later he was sorry that he killed a wolf.

5. Ethics for Wholes

How does ecocentrism differ from other kinds of naturocentrism? Ecocentrism is easy to distinguish from zoocentrism, which stresses the well-being of individual sentient beings, albeit it is not easy to differentiate from biocentrism. In the broader understanding of life, rocks and mountains are regarded as living beings. Are ecosystems living entities, as, for instance, Johnson claims? We may mean by biocentrism individual-centered views, like Taylor's biocentric theory. Biocentrism is individualistic: biological life is a distinctive character of individuals. Biological life necessarily refers to individual living beings. In biocentrism, as I define it, life (bios) is the most central concept. Ecocentrism is primary holistic. Ecocentrism encompasses the topics of wholes like species and ecosystems (anthropocentric and naturocentric concerns for ecological crisis and the well-being of ecosystems), or more concretely, rivers and mountains, and non-living particles that create these wholes (physiocentrism). Beauty and the discussion of diversity in nature which do refer to concrete wholes such as rivers and mountains, belong to the domain of ecocentrism. We need have no respect for individual rocks and stones, but since these form wholenesses like hills and mountains, we do need the attitude of respect for them.

Ethics for space, the universe, and the cosmos follows from the issue of ecocentrism. Does the universe have intrinsic value? Do other planets? The planet Earth is unique in its life-support systems. Do other planets remain without value? I restrict ethical extensionism to the Earth, all individuals, things, states, and wholes which we can benefit or harm. Where we can harm or benefit something, we cannot have an ethically indifferent attitude. If I have the opportunity to harm something by my action, I have to decide whether I am morally allowed to do this action. Since we do not yet have abilities to harm other planets, I restrict my notion of ethical extensionism to the Earth. The discussion of whether Mars has intrinsic value, although theoretically an interesting question, is not a crucial problem at present in practice.

We may argue that in the naturocentric context, the concept of nature is too big, including the whole order of nature. We should respect nothing less than nature as a whole, in which we are integral parts.[28] However, according to zoocentric or biocentric thinkers like Regan and Taylor, the main concern is for individual animals and plants. We are wrong in supposing that nature concern is always for the environment in the broadest sense (the biosphere, the order of things.) Only physiocentrism sees the question as of the environment in its largest meaning. The problem of physiocentrism include, how we can make sense in speaking about "nature as a whole" and how we can discover intrinsic value in this whole in a meaningful sense. Another problem is the human place in the nature. I am part of nature, but I am a highly separated part. How can we put these two sides together? In reality, people do have power over nature. On the naturocentric perspective, we should not accept that they have absolute moral priority over nature.

Environmental philosophers hesitate at this kind of ethical extension, at the level of ecosystems and nature as a whole, because of the common-sense idea of the significance of individuals in ethics. Many philosophers suppose that only individual beings can be benefitted or harmed. This dogma is left as a result of the extension of human-centered ethics to the non-human world. Regan and Taylor have based their environmental philosophical theories on grounds in which the standard for having intrinsic value is an individual human being. Our Western attitude has been ethics for individuals, not ethics for wholes. Accordingly animal suffering is taken morally seriously into account by moral philosophers who hesitate to speak of ethics for species and ecosystems. Yet the language of ethics for species and ecosystems is easily accepted among biologists and environmental scientists who may hesitate to attribute intrinsic value to animal life. Although ethical extensionism is beautiful in its logic, beginning from our duties to other people through animals and living beings to ecosystems, in practice it does not function.

Anthropocentrism, zoocentrism, biocentrism, and ecocentrism may have their defenders and activists who have no ideas of ethical extensionism, of the logic leading from human welfare to animal well-being, from animal well-being to caring for the well-being of all life forms, or to caring for the well-being of ecosystems and finally the whole earth (geocentrism or physiocentrism). Your attitude can be anthropocentric and ecocentric, if it is indifferent to animal suffering. Moral attitudes do not follow the logic of ethical extensionism. A person may have empathy with other people, and plants, and ecosystems. The conservation of nature is often not naturocentric at all in its underlying philosophy. Many non-naturocentric human interests occur in nature: economic, ecological, scientific, religious, cultural, aesthetic, and leisure interests. The most radical animal activists are only interested in the question of animal welfare, and sometimes not even interested in human welfare. Some activists on behalf of endangered species do not care for individual life forms, not even for human life. Some may have strong empathy for other people and the environment as a whole while having an indifferent attitude to animal suffering.

Human-based environmentalism (concern for the environmental crisis)	Naturocentric positions (concern for the well-being of animals or nature)
Ecohumanism Ecosocialism Ecofeminism Ecomysticism	Zoocentrism (animals) Biocentrism (life) Ecocentrism (ecosystems) Physiocentrism (nature as a whole)

In anthropocentrism, human intrinsic value is always superior to non-human intrinsic value. Being non-anthropocentric does not necessarily mean being naturocentric. You may be a non-anthropocentric thinker, not naturocentric. For example, all duties may be ascribable to God, not to human or non-human beings on the earth. Another possibility is to understand the importance of the ecological crisis without any concern for the well-being of non-human animals and nature. The main eco-terms refer to the ecological crisis or other environmental problematics. To be environmentalistic does not necessarily imply being naturocentric or being concerned about the welfare of non-human animals and nature. We must distinguish environmental attitudes concerning the ecological crisis from a naturocentric attitude that stresses the well-being of non-human animals and nature. "Environmentalism" refers to the ideology of ecology as a movement that stresses the ecological crisis. For example, ecofeminism is a synthesis of ecological and feminist thinking, because it recognizes the importance of the ecological movement and of feminism. From a biological point of view, endangered species have a special value. The forest gets value from its endangered or rare species. This is a biologically-based conservation attitude. Those with a strongly biological attitude to nature need not care about human well-being at all. Their interests can be purely biological or scientific without any reference to human welfare; indeed, science can run against some human interest. Naturocentric interests are human interests in nature based on the well-being of animals or nature, and not primarily on human well-being. Yet nature can be preserved on several anthropocentric grounds, for future generations, for scientific purposes, and for medical or agricultural reservoirs, even without any concern about the ecological crisis or the well-being of non-human nature.

6. Natural Beauty

Leopold argued that our actions are right if they support the integrity, stability, and beauty of nature. Natural beauty is a prominent candidate for ecocentric value. Beauty is a special kind of value. People value aesthetic properties without regard to use value. Beauty in nature is thus ecocentric, value-centered on nature, and at the same time anthropogenic. Nature is beauty for people. But is it beauty in itself? Intrinsic natural beauty (a natural thing-being-beauty-in-itself) is found in nature; people have not created it.

I develop further the notion of natural intrinsic beauty, by extending the origin of beauty to the non-human world. Finally, I propose a model for naturalistic aesthetics that takes seriously the possibility of a non-human-based-aesthetics, trying to discover zoogenic, biogenic, and ecogenic beauty in nature. In this sense, natural beauty is an ontological value, value generated by nature.

A crucial issue is an old philosophical problem: is beauty only in human eyes? Anthropocentric positions ascribe intrinsic value to natural beauty. Such value is ascribed to nature, not found there. The beauty of natural areas and plant and animal species can be seen as necessary for human well-being. Natural beauty thus has instrumental value for people. In addition to this, natural beauty or nature's aesthetic values can be regarded as a part of nature's anthropocentric intrinsic values, that is, anthropogenic value. Evidently, people have appreciated beauty for its own sake in the natural world regardless of nature's use value for human purposes.[29]

We need to distinguish aesthetic perceptions from aesthetic objects, or human abilities from nature's properties.[30] Aesthetic abilities are in beholders, while aesthetic properties are objectively in natural things. Beauty emerges from nature in that the presence of a beautiful object is necessary and fundamental for it to be aesthetically valued for its beauty. The sense of beauty is in the mind, while the object, say, wildness, that is sensed and that generates the experience of beauty, is not. According to this, beauty can be understood in relational terms. Beauty is not in objects or not in subjects. It is in relation to both.

"Beauty" is more anthropocentric than "good." You might claim that aesthetic experiences center on people since animals and plants have no sense of beauty.[31] Could "biological beauty" exists in nature as well as "biological goodness"? Animals do have some sense of beauty. The nests of birds are decorated with colored objects such as fruits, flowers, shells, and bones. The color preference may vary within the species and individual birds.[32] The evidence of sex-specific color patterns in many animals leaves open the possibility that beauty could be generated by non-human animals, an issue to which I return in the end of this chapter.

A. The Creation of Beauty

"A beautiful sense object is beautiful, not because it is sensory, but because it illustrates harmony, balance, development, unity in variety, or some other aesthetic principle, or expresses the artist's or the observer's soul," says Edgar S. Brightman.[33] The idea of beauty in nature is related to such concepts as unity in variety, harmony, wildness, and naturalness. We find support for this kind of idea from the concept of intrinsic value as developed by Robert S. Hartman. In "The Nature of Valuation," he describes the creation of intrinsic beauty as the unity of the picture and its painter.[34] A painter, when in the state of intrinsic valuation, gives her or his soul into the picture. The genius artists put their whole personalities into their works. In the state of extrinsic valuation, the painter paints with colors and painting techniques; the result is a picture. In the creation of intrinsic beauty, the picture is an extension of the painter's personality. This creates a new whole, a unique work of art. Hartman calls this the organic unity of artist and work of art. Here "organic" means the same kind of organic unity Moore has in mind, the new whole which cannot be reduced to its parts. Organic is roughly a synonym for holistic. The principle of organic unities refers to the idea that the value of a whole is not the sum of the values of its parts.[35]

The idea of Hartman's intrinsic value may be extended to the natural world in that no human being is necessary for the intrinsic value of the organic unities of non-human nature. Yet Hartman's view is strongly anthropocentric, although it gives value to people and the whole process. Hartman's view of intrinsic value may be difficult to apply to the non-human world, since it presumes the existence of human beings as the creators of intrinsic value or of beauty in that an artist puts her or his life and soul into a picture. The intrinsic value of the picture is then the extension of the artist's personality.

This kind of idea of intrinsic valuation can be found in transcendentalism. According to Ralph Waldo Emerson (*Nature*) and Henry David Thoreau (*Walking*), the presence of human beings in nature can transcend the non-human world by bringing forth its intrinsic value and beauty.[36] The idea can be understood in much the same way as Hartman describes the creation of intrinsic value. Nature-lovers put their lives and souls into the preservation of wilderness. The intrinsic value of nature is the extension of nature-lovers who give their souls into nature. Nature becomes part of this newly created whole which contains more than nature and nature-lovers. This process creates the intrinsic value and the beauty of nature. The existence of nature to be loved and those to love it are necessary to create this kind of intrinsic valuation of nature. This is a kind of subjective and anthropocentric valuation. It may be called an interactive and weak anthropocentric valuation. Both the valuer and the valued are necessary in the creation of intrinsic value in the language of intrinsic valuation.

A more objective theory of value argues that not everything can be valued. The intrinsic value-properties of what is valued determine the value-event. Not everything can be good and beautiful in nature. Some qualitative properties and the character of beauty in nature are discoverable by human beings.

B. Anthropogenic Beauty in Nature

For Eugene Hargrove the beauty of nature is a value-laden natural state.[37] He has defended nature's anthropocentric intrinsic values, especially nature's aesthetic values, as having intrinsic value. He remarks that people have valued and still value natural beauty for its own sake. This kind of valuing is based on anthropocentric grounds. Natural beauty can be understood as a part of the total good in the world. Thus, its existence should be promoted and preserved. Hargrove's ontological argument for nature conservation based on the aesthetic intrinsic value of nature is as follows:

(1) Human beings have a duty to promote and preserve the existence of good in the world.
(2) Beauty, both human and natural, is part of good.
(3) Thus, natural beauty is worthy of being promoted and preserved.
(4) The existence of nature precedes natural beauty; natural beauty is contingent upon nature's physical existence.
(5) Thus, the need to preserve nature is greater than the need to preserve human art.[38]

On this ground, Hargrove claims that natural objects take precedence over our obligation to protect works of art. We have a stronger need to preserve natural beauty than human art. From this point, Hargrove's view is not precisely anthropocentric, although he claims it is. A non-anthropocentric, naturocentric thinker could agree with this view that recognizes the primacy of the natural world to non-human art. Natural beauty comes to have moral significance for us. Needless destruction of natural beauty is wrong. Beauty and goodness resemble each other; beautiful things or states can be said to be good things or states. Thus, beauty cannot be completely distinguished from what is good. The term "beauty," like "goodness," is value-laden. To say that a natural area is beautiful but that if we destroy the area it does not matter, is contradictory.

Support for the intrinsic value of natural beauty can be found in Moore who argued that goodness and beauty are intrinsic rather than extrinsic kinds of value.[39] The goodness or beauty of an object depends solely on the intrinsic nature of the object in question.[40] Accordingly, the beauty or goodness of a thing depends on the intrinsic nature of the thing. No extrinsic reason affects these values. Another feature of this kind of intrinsic value such as beauty and goodness, is that if one thing possesses such a value, other things exactly like it possess this value, too.[41] This idea has been called "supervenience." Supervenience means that, for example, the aesthetic properties of a thing can be said to supervene on its physical properties in the sense that if two objects are alike in all physical respects they must be alike in all aesthetic respects.[42] Or as in Moore's example, intrinsic value as an axiological property supervenes on natural properties in this sense.

C. The Naturistic Model of Beauty

The beauty issue does not necessarily touch the main ecological and reverence-for-life issues. Conversely, we may propose just ecological and ethical criteria for assessing what is beauty. Modern environmental aesthetics began in the 1960s in relation to the environmental movement. Ecological values such as environmental health, life values, dynamic processes, and the value of naturalness are investigated in environmental aesthetics. The ethical point of the harmony of people and nature is taken into aesthetic account. Beauty involves many different aspects, which form the totality of beauty.[43]

The basic values of ecological ethics are reverence for life, responsibility for our lives, frugality as a precondition of inner beauty, the pursuit of wisdom, self-actualization, and understanding of the heritage of life and of evolution.[44] The need to preserve and enhance the environment around us is an ecological imperative for environmental aesthetics.[45] Illness, suffering, or badness are not the general characteristics of environmental art. Animal activists have questioned the fashion for furs wearing on ethical grounds, by claiming that the beauty of clothes is no real beauty if it is based on animal suffering. A purpose of environmental art could be to reveal suffering to the general audience, thus helping animals.

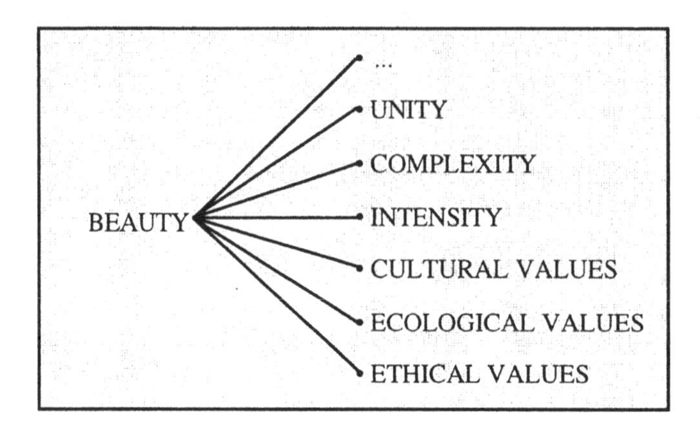

Finally, I apply my naturistic model to the beauty issue. What do we mean by "intrinsic natural beauty"? First, the term "intrinsic" refers to the inner or inherited value in nature. "Intrinsic" means a non-human based value, value in nature. My attempt is to develop a non-anthropocentric conception of value. Although beauty is commonly regarded as the most anthropocentric of all values, related to the human aesthetic experiences, my task is to revalue and reconceptualize "natural beauty" as a non-anthropocentric value in nature, or more exactly, as a naturogenic value in nature, value-rooted or generated originally in the non-human world. Second, by the term "natural" I refer to the three topics (zoo, bio, physis) examined throughout this book. A basic proposal for environmental aesthetics is that we should develop (1) animal aesthetics, (2) life aesthetics, and (3) ecosystem-based-aesthetics. We need these basic concepts for a natur(al)istic aesthetics to distinguish it from the human-based aesthetics. Aesthetic features are so extremely common in the non-human world that we should wonder why nature's aesthetics has been an ignored issue in the field of environmental aesthetics. Third, I understand the term "beauty" to refer to all aspects of aesthetics: visual, musical, and taste. A zoogenic, animal-rooted beauty, appears in the varieties of animal activities: singing birds, dancing foxes, howling wolves, the extraordinary sounds of whales and dolphins. These animals express themselves and communicate with each other by their musical sounds and moves. I believe that beauty, including the elements of harmony, feelings, or power, is significant for these animals personally and in a social sense. Zoogenic aesthetics will get its empirical material from animal architects and designers (decorated nests, sexual body colors). Some well-trained dogs are real exhibitionists who enjoy presenting themselves at dog shows.

Life aesthetics is a more difficult topic to study. I do not believe that the colors of flowers are for the sake of us or for animals. Could aesthetics have an certain inner significance for plants and trees? If we can speak meaningfully about the good of plants, in Taylor's terms, can we develop the topic and terminology of the beauty of plants? This would be called "biogenic beauty" in nature, beauty rooted in living organisms. The next step is to develop an ecosystem-based-aesthetics, which discovers beauty in the non-human world and expands the holistic language of natur(al)istic aesthetics.

I have offered proposals of how the threefold terminology of naturogenic aesthetics (zoogenic, biogenic, and ecogenic beauty) can be applied to an environmental aesthetics. I encourage extending discussion from anthropocentric to naturocentric aesthetics.

PART III

THREE BASIC ISSUES

Seven

THE ORIGIN OF VALUE

Where do values come from? Two basic answers are: from culture or from nature. The cultural view of value explains values as human constructs. It can be classified as cultural constructivism. Typically, values are human aims or targets. Basic human values include health, work, home, friends, and hobbies. Money is an instrumental value, while good health is an intrinsic value, valuable as an end-value. We may speculate whether all values are cultural and historical human constructions. A possibility of non-human value remains. Thus, cultural constructivism is not a sufficient answer to the problem of anthropocentric and naturocentric value, a division which is real and vital in the philosophy of nature conservation. The basic questions remain: What is value? What is the origin of value? And what is this thing called the intrinsic value of nature?

My aim is to discover values in natural entities. A pure value-fact division is passé, as recent arguments suggest. "Facts" are sometimes created about states of affairs concerning human existence and well-being.[1] Facts have no more objective, human-independent nature than do values. Scientific facts are theory-laden, as is well-known by philosophers of science. We live in the world impregnated by values. Facts and values are regularly mixed in practical life: our practical reasoning is based on some premises which are entangled with values. The difference between facts and values may exist but this is of no practical significance.[2] In nature, we can describe facts and values as inseparably co-evolving. Nature is the bearer of value, and in addition, it is the origin of value. Cultural and natural things and states, being value-laden and fact-laden contain values and facts. A view of the intrinsic value of nature is based on the notion of the good of each living being, which is value-laden "fact" of nature.

The languages of fact and of value are human languages. Facts are no more real than are values. We can use biological terms in describing, understanding, and explaining the natural world, while the philosophy of nature uses the philosophical terms. We have two different languages of nature: the language of biological facts and the ethical language of values. Nature philosophers will be those future specialists who help people to learn values in nature, similarly to the way in which biologists can tell what kinds of species exist, or what the biological functions are in nature. We can perceive values in nature, if we have learnt to perceive them, just as we have learnt to perceive biological or ecological facts. The topic of teaching values belongs to the environmental educators. By valuing nature or parts of it, we are answering the question of what nature is. Perceiving values in nature means perceiving nature in the ethical way. Nature is worth being protected. Nature is a value concept, just as is a sentient animal, a flourishing tree, or a beautiful landscape.

We should believe in some kind of natural realism according to which real matters exist independently of human beings. Such matters as the good of living beings, or animal suffering occur in this reality. I suggest that this be called the natural constructivism of value which we should distinguish from cultural constructivism. Both cultural, human-based history, and natural, nature-based history create the complex reality of the world of value.

1. Subjective Valuing and Objective Values

The epistemology of value is subjective, objective, or relative. Value-subjectivists claim that nature has no value on its own. It only has value as it enters into relationship with valuing subjects, human or otherwise. Values are constructed inside valuers. The quality of objects does not affect our valuation. For value-objectivists, value is independent of the opinions, valuations, or interest of subjects. Value depends on an object's intrinsic, qualitative nature. If the object is removed from its context and relocated elsewhere, its value is unaffected, since it inheres in the object. Relative value can be developed by interaction between us and natural beings, between valuers and value-objects, albeit it can exist between two natural objects. A plant or an animal is in value-relation to other beings and to the environment.

We need to distinguish between ethical and ontological issues. You can be anthropocentric in your ethics, and still defend, say, a zoogenic theory of the ontology of value (for example, a kind of evolutionary model of value). Or you can be an ethical extensionist, while thinking that all values are human-generated. You can attribute intrinsic value to a non-human thing in the sense that you appreciate it for its own sake, as an end-value. Or you can think that the planet Earth has intrinsic value because you appreciate it as intrinsically valuable. You can claim that people generally should value this planet instrumentally as well as in terms of ends-value.

We can attribute anthropogenic value to nature. And we can discover naturogenic value already present. "Value" does not mean the same as "valued by someone." The fundamental question of the ontology of value is: "Are things valuable because we value them or do we value them because they are valuable?"[3] The basic axiological assumption is that we value something because of its value. The value of an object gives us reason to value that object. Thus, having value, being valuable, constitutes what will be valued by someone. Accordingly, objective and subjective values can be defined as follows: Value is objective if its existence and nature is independent of a subject. Value is subjective if it owes its existence, its sense, or its validity, to the feelings or attitudes of the subject.[4]

Intrinsic value	Extrinsic value
Value in itself	Value for something
Objective value	Subjective value
The existence of value is independent of a subject	The existence of value depends on a subject
The value of objects	The value of subjects
Objective values	Subjective valuing
The intrinsic qualities of objects	The interests of subjects

An example of the objectivist argument is that the quality of wine is not only present in our physiological and psychological taste, but it also inheres in wine. We should deny that all values are connected to people: "If we make man the measure of aesthetic value and moral law, it would appear that there would be, strictly speaking, neither good taste nor morality."[5] Things can be valuable or non-valuable for the reason that we human beings decide to judge them as valuable or non-valuable. But if the values of cheap and fine wine are the values we give them without reference to wine's value-properties (which are objectively valuable), no agreement could occur as to which wine is better than another. Moreover, no less is true that agreement is greater than disagreement.[6] In the subjective value-context, all moral education and other education would be impossible. If values are nothing more than assigned valuations regardless of appropriate or inappropriate properties actually present, no right or wrong exists on moral issues.

According to Risieri Frondizi, subjectivism is right in asserting that value cannot be entirely divorced from valuation, but it errs in reducing value to valuation.[7] Frondizi defends on interactive theory of value. Value is not in subjects nor in objects but arises in interaction between them. Any object's real value-properties determine the valuing subject's valuation-processes, but cannot be actualized for us human beings without our conscious states, which must be able to value some things. This view is opposed to the strong objective intrinsic-value view, and defends the relational-value view, according to which no isolated intrinsic values exist. Values are relational and plural in their nature.

The confusion between values and valuers may be called the subjectivistic fallacy, or the anthropocentric fallacy.[8] We evaluate the world as human beings, but a distinction must be made between the valuing subject and the object or content that is valued.[9] As Frondizi has aptly said, "a distinction should be made between valuation and value. Value is prior to valuation. If there were no values, what would we evaluate? To confuse valuation with value is like confusing perception with the object perceived."[10] An example of objective value is human reason. Even value-subjectivists have to accept that at least reason is of intrinsic value. They have to agree with value in reason, because their argument is based upon reason. The argument is that if value is fully relative or subjective in its nature, value-subjectivists could not argue for the truth of their position.[11]

We could argue that the value of human reason is not an objective value but just a value that we attribute to it, but the problem remains, why do we give it this value? Thus, the basic axiological question is, do we value something because of the value it already has? Some qualitative properties of a thing make it valuable. The value-properties of an object gives us reason to value the object.

2. Weak and Strong Objective Value

We need to distinguish between a weak and strong objective value. By weak objective value, I mean that the value at stake is objective in that what is my personal opinion does not determine my valuation, but my valuation is grounded by well-reasoned arguments and by acceptable general or ethical rules. Valuation is impersonal and intersubjective. The strong objective-value view refers to the view that value is ontologically independent of human valuation.

Intrinsic value	*Ontology*	*Ethics*
Weak objective value	Dependent on human valuation (relational value)	Impersonal, inter-subjective
Strong objective value	Independent on any valuation (isolated value)	Attitude of respect for the world

In philosophy, if we take values seriously, to regard something as having value requires us to have well-grounded arguments. Each value should be reasoned by asking, "Is this a value? Why this is a value?" If I feel that I want to kill somebody, my feeling does not justify murder. If I fall in love, this is no reason to marry. Emotions and feelings should be reasoned and analyzed. And analyzing values should be separated from acting on values. Human values are not the same as our changing interests but what our best interests ideally are. In this sense, all human values should be weak objective values, human-based but well-reasoned values. The question of natural value is more diverse. Nature does not argue for its values; it acts them. Value in nature cannot be a weak form; it is by necessity a strong objective value. At the bottom are basic biological and survival values.

3. A Naturistic Model

If we say that things have intrinsic value in the sense of autonomous, objective value, then we have to pursue further the question of the naturogenic value-theory, according to which animals, plants, and ecosystems have value in themselves. In the following sections, I develop and characterize my views of zoogenic, biogenic, and ecogenic value, to distinguish them from zoocentric, biocentric, and ecocentric value.

Anthropogenic intrinsic values in nature	*Naturogenic value*
Weak anthropocentric intrinsic value (Hargrove)	The good of living beings (Taylor)
Truncated intrinsic value (Callicott)	Inherent worth (Taylor)
	Inherent value (Regan)
Indexical intrinsic value (Elliot)	
Existence value (Randall)	Natural value (Rolston)
Naturocentric values: zoocentric, biocentric, and ecocentric values (Vilkka)	Zoogenic, biogenic, and ecogenic value (Vilkka)

A. Zoocentric and Zoogenic Value

The terms "suffering" and "well-being" link facts to values. The suffering of an animal is an empirical fact of the animal and a value-term. Animals try to avoid suffering if they can. From the zoogenic perspective, the well-being of animals has intrinsic value. Intrinsic value is found in the well-being of animals, and not only in one's attitudes. Suffering is intrinsically bad for animals, while well-being is intrinsically good for them. Animal suffering is bad for animals, not always for people. According to the zoogenic view, at question is the suffering of the animal, not our suffering. People can gain knowledge of animal suffering or well-being from within their value consciousness. The zoocentric and zoogenic value of animals can be understood as follows. The value of animals can mean

1. Their personal or individual value to other beings (the value of love and companionship) (*zoocentric value*).
2. Their value on the grounds of their well-being from their own point of view (*zoogenic value*).
3. Their self-value: an animal values its own life and interests (*zoogenic value*).
4. Their value on the grounds of their universal qualitative nature as a sentient being (*zoogenic value*).

The value of an animal depends on both

(1) its animal-centered relations to other entities and
(2) its intrinsic qualitative nature.

In the first case (1), the zoocentric value of an animal can mean its value-relation to other beings (the value of love and companionship). Its value to people can be related to its self-value. An animal values its own life and interests, which we human beings respect. Animals care about their own lives, normally having a strong interest in not being killed. In the second case (2), an animal has zoogenic value in virtue of its intrinsic qualitative nature. This can be understood in terms of animal well-being or in terms of an animal as a sentient being. The survival value of conscious states is presumably developed in the course of evolution. Consciousness has survival and adaptive value for those who are conscious. Hence, the value of a conscious state is not dependent on anybody valuing it, although conscious states are intrinsic goods in nature. Conscious states would not have evolved if they had not been intrinsically good, in evolutionary value, for their possessors.

The value of an animal does not depend on what others think about it, say, whether it is a poor or excellent dog. Each animal has a quality of being as such, which is more than our valuing it as we people perceive it. Animals have their own distinctive intrinsic values as the animals they are. Another being with an integrity, a dignity, and value of its own, demands our respect. Such an other being can be a squirrel, a cow, a rat, as well as a human being. The zoogenic-value view involves the idea that animals derive their value from their well-being, and this well-being is determined from the animals' point of view. A zoocentric value is at stake, when we claim that we should love or respect animals.

A ZOOISTIC MODEL	
The zoocentric issue	The zoogenic issue
Values centered on animals (love and respect for animals)	Animals as an origin of value (animal wellbeing)
ETHICAL/VALUATIONAL ISSUES	ONTOLOGICAL/EMPIRICAL ISSUES

As a theory of animal values, a zooistic model, is divided into (1) zoocentric and (2) zoogenic issues. Similarly we could figure out the whole naturistic model with its threefold rubric (zoo, bio, eco).

B. Biocentric and Biogenic Value

Many environmental philosophers have restricted intrinsic value to sentient animals. Could only conscious states or experiential states such as pleasure and happiness have intrinsic value? They do, but they are not the only ones. I argue for a pluralistic value-theory in which not only conscious states have intrinsic value, but intrinsic values emerge from many different things, beings, and states. In the following I give two examples of the intrinsic value of life: (1) the principle of biodiversity in nature (biocentric value), and (2) the good of living beings (biogenic value).

1. The notion of the diversity of life has recently become one of the most powerful ideas of nature preservation. Even if some life forms occur on other planets, our planet undoubtedly remains unique. Respect for life is basically respect for life forms, our planet's life forms, not for individual lives. The unique character of life is its ability to vary, to form new and different life forms. The intrinsic value of the flourishing of diverse life forms is a basic principle in deep ecology. Biodiversity is valuable. The varieties of life-forms are to be protected and preserved. Biodiversity is a biological fact that can be measured, and it is a value-laden conception.

2. According to biocentrism, human and animal good must be taken into moral account and the good of all living beings. Taylor's basic value is the inherent worth of living beings. If an entity has a good of its own, and if it is better for the entity that this good be realized than not, then the entity has inherent worth. The good of living beings and their inherent worth is the fundamental value for the attitude of respect for nature. We can separate the good of living beings from values relative to

any other beings and from merits. Those who have the good of their own kind have it in their own qualitative nature and it belongs to them inherently.

The concept of life is a value-concept, a qualitative concept. Life is a biological fact and qualitative, intrinsic property of a living being. Life is axiologically more than mere being (existing). Vitality is a universal, general intrinsic value, more basic than consciousness as an intrinsic value. A living being is not merely quantitatively something more than a non-living entity. It has qualitative features. X-being-alive compared to Y-being-inanimate is not X having abundantly something which Y has, but X has a novel qualitative and distinctive value-property. I call this characteristics of life the value intrinsic life, this intrinsic value of life in terms of inherited or inborn value. This is the biogenic intrinsic value generated in the variety of life forms.

C. Ecocentric and Ecogenic Value

Does only possessing (biological) life make something worthy to be respected, loved, and preserved, while all non-living things have only instrumental value or are otherwise valueless? The position which restricts moral standing to living beings is mistaken. By the ecocentric theory of value, I mean the view that nature as ecosystems, mountains, mires, and rivers has intrinsic value. The notion of the biotic community is a key concept of ecocentrism. It extends the idea of community to include not only living individuals as animals and plants, but land, water, and air, as well as forests, rivers, mountains. From the holistic point of view, ecosystems and the earth are the source of different kinds of values and of the variety of intrinsic values. The earth is value-laden, not barren of values. In this holistic perspective, the ecosystems on the earth are value-generating systems. They are able to generate the variety of values. Hence, nature is valuable for us and able to produce values in itself.

Is this the fallacy of origins? If something generates values, then that it has value in itself does not follow. Also, if something has ability to generate value, it does not follow that it is in itself valueless. If X has ability to generate value, that might be a sufficient reason for us to value X for its own sake. Ecosystems do have intrinsic value for us, in terms of ends-value, or for their own sake. Indeed, Laura Westra has defended the position that recognizes intrinsic value in life-supporting wholes.[12] Do ecosystems have intrinsic value in terms of their well-being? Johnson argued so.[13] I agree with him that ecosystems can be destroyed or preserved as having a good of their kinds.

Does a genic value, ecogenic value, exist in the ecosystem? Ecosystems able to generate value are valuable in themselves because of their intrinsic qualities. Systemic value refers to the level of ecosystems. We human beings can value nature as a whole for many reasons. The beauty of nature is a prominent case of ecogenic value.

4. A Genealogy of Value

A basic difference between anthropogenic and naturogenic value-theory concerns the ontology of values. According to anthropogenic theories, people can value animals, plants, and ecosystems instrumentally and "intrinsically." This "intrinsically" refers to the class of ends-value, that is, things or states valued for their own sake, which belongs to the class of values which are attributed, ascribed, or given to objects. If we say that things have intrinsic value in themselves in the sense of strong, objective value, then at question is the naturogenic valuetheory, according to which animals, plants, and ecosystems have naturogenic value in themselves.

THE FORMS OF INTRINSIC VALUE IN NATURE

1. The types of intrinsic value in nature

A. Zoocentric and zoogenic value
B. Biocentric and biogenic value
C. Ecocentric and ecogenic value

2. The specific ideals of intrinsic value

A. Well-being
B. Biodiversity
C. Beauty

3. The generic levels of intrinsic value

A. Consciousness
B. Life
C. Ecosystems

The three fundamental levels of values exist: the value of consciousness, of livingness, and of wholeness. From the naturogenic point of view, these are inborn, inherited values in sentient beings, in living beings, and in nature as a whole. X being an animal, a plant, or an ecosystem, X's intrinsic value is based on three elements which are related to each other:

(1) X having naturocentric and naturogenic value (the naturistic types of intrinsic value, X having zoocentric or zoogenic value, biocentric or biogenic value, ecocentric or ecogenic value).
(2) X having well-being, X increasing beauty, or biodiversity in the world (the specific ideals of intrinsic value, consequently on the levels of zoogenic, biocentric, and ecocentric value).
(3) X's generic nature as a sentient being, as a living being, or as a whole (the naturogenic levels of value: zoogenic, biogenic, and ecogenic).

The generic levels of value	*The specific ideals of value*	*The types of value*
Consciousness	Well-being Love animals!	Zoogenic value Zoocentric value
Life	A good of living beings Biodiversity	Biogenic value Biocentric value
Ecosystems	Nature able to generate values Natural beauty	Ecogenic value Ecocentric value

In this theory, naturogenic value in nature is its qualitative property, which is an inborn feature in nature. Naturogenic value is a natural qualitative property generated in animals, plants, and ecosystems. A sentient animal has that qualitative value intrinsic in all conscious beings which non-conscious beings do not have, that is, zoogenic value. A living being has a qualitative value intrinsic or inborn in it, which the non-living entity does not have, that is, biogenic value. Nature as a whole, like the forest ecosystem, contains a qualitative intrinsic value, that is, ecogenic value.

According to weak anthropocentrism, we people can attribute intrinsic values to non-human beings. We ascribe this value to some non-humans (anthropogenic intrinsic value). But from the naturistic point of view, which I defend, the values in nature consist both of anthropogenic and naturogenic values. Naturogenic values are intrinsic qualitative properties in nature which are there independently of any human attribution. Zoogenic, biogenic, and ecogenic values are the values inhering in animals and in nature independent of the valuation by human beings. Such values exist whether people perceive these intrinsic values or not.

I have defended the naturocentrist position in its ethical extensionist sense: animal suffering should be taken into moral account as well as the well-being of non-sentient nature. I have questioned both the ethical and ontological view that the non-human world has merely anthropogenic and anthropocentric value. But any criticism of the assumption of merely anthropocentric values needs a more positive answer to establish what values are beyond anthropogenic and anthropocentric values. On ethical perspective, I developed my naturocentric argumentation in which different kinds of entities belong to the scope of our morality on different grounds: animals as sentient beings (zoocentrism); living beings because of the value of life (biocentrism); the whole planet Earth because of its unique life-support systems (ecocentrism). From the ontological point of view, I developed an answer in terms of naturogenic value, the value which a non-human entity has because of its inherent qualitative characteristics. Naturogenic values are intrinsic, qualitative properties generated in animals, living beings, and ecosystems. These are able to create values. Thus, they are value-laden and not intrinsically barren of values. This theory of the intrinsic value of nature is based on an ethical extensionist and pluralistic theory of value.

5. Existence-Value

The theory of the intrinsic value of nature is a challenge to the strong anthropocentric positions. In environmental economics, three basic value of nature, which can be measured by terms of money, are use-value, option-value, and existence-value.[14] By use-value, economists mean marketing value for different kinds of nature's products, which demands utilizing nature mechanically or as a raw-material. Option-value means value a natural area will have in future. Existence-value has been developed to indicate people's willingness to preserve nature as such, without utilizing it, and their willingness to pay for this. Direct willingness to pay can be measured by asking people how much money they would give for the existence of an animal species or a natural area (this is called "Contingent Valuation Method," CVM). The indirect willingness of people to pay for the existence of nature can be measured by investigating their travel costs to reach a wilderness area.

Alan Randall[15] argues that "total value" takes into account use-value, option-value (amenities in the future), and existence-value in nature. Existence-value of nature is a novel and radical notion in environmental economics. Existence-value is roughly understood as a synonym for intrinsic value in terms of ends-value. My criticism is that this total-value view is exclusively an anthropocentric and instrumentalist view. Animals, nature, and future generations in themselves are not taken into account in this total-value context, since values are expressed as human willingness to pay. We would be more correct in speaking about the framework of human value in terms of willingness to pay. This is the view according to which human values can be accounted for in the last analysis in terms of money.

The value of human existence cannot be expressed in terms of money, and the same should be true of nature's existence values. Nature as good in itself must be distinguished from the question: how much would people pay for the existence of a natural area? The existence-value of animals and nature for people is anthropogenic value, while nature's existence-value is naturogenic value. The economic existence-value of nature is a kind of intrinsic value, a value for people. "Total value" takes into account no naturogenic value in nature. It ignores the possibility of naturogenic values in nature, since according to it all values are in relation to human subjects. Thus, the so-called total-value view ignores the major part of the varieties of the intrinsic value of nature. To my mind, environmental economists have the crucial task of developing models which take nature's intrinsic values into the economic accounting. The idea of existence-value is a good start.

Eight

ANTHROPOCENTRISM AND THE PROBLEM OF PRIORITIES

Some philosophers are concerned to find solutions to environmental crises on anthropocentric grounds. Their aim is to develop anthropocentric theories and principles to justify environmental ethics without the notion of the intrinsic value of nature. I call these views strong anthropocentrism to distinguish them from those anthropocentric theories which approve our assigning intrinsic value to nature. The view according to which the non-human world has intrinsic value generated and assigned by people is here called weak anthropocentrism. We value from a human perspective, but it does not follow that we value necessarily in terms of human instrumental interests.

Some anthropocentrists have claimed that nature is a value-in-itself, because it is an inseparable part of human existence.[1] Nature cannot be an instrument purely, because our existence is absolutely dependent on it. If we have intrinsic value, so does nature, because it is an inseparable part of us. This would be called the oneness argument: the whole of nature reveals itself in a human being. Nature and people cannot be separated from each other; they form a totality. Living in harmony with nature means living in such a way that we really support our own goals. We have our inner nature, nature-in-us, but we are a separated part of nature. We are individuals in wholeness. We need to take outside-nature into account. We must protect soil to produce crops. There must be fresh water in the city. This forms the significance argument: the well-being of outside-nature is enormously significant for our lives. Is this outside-nature of intrinsic value? Logically, if something is enormously significant for us, it does not follow from this that it has intrinsic value. This view may assign to outside-nature value over and above anthropogenic value. This kind of view may defend nature's instrumental value, which is not created by people. It defends nature's instrumental naturogenic value, that is, instrumental values generated by nature. The most significant of intrinsic values are life-supporting values in nature.

Many philosophers, who designate their philosophy on anthropocentric, accept the view of the intrinsic value of nature in terms of human valuation. Eugene C. Hargrove speaks about anthropocentric intrinsic-values. The object of our valuation does not transpose our ethics to a non-anthropocentric kind. If we follow Hargrove, natural beauty is an anthropocentric intrinsic-value. Valuing natural beauty for-its-own-sake refers to the anthropocentric intrinsic-value. But, then, what would be a biocentric intrinsic-value? If I value life-forms for their own sake, this is a type of anthropocentric valuation. Another possibility is to define the type of valuation in accordance with the objects of valuation. If the object is a human being, then our valuation is anthropocentric. If the object is an animal, our valuation is zoocentric. If we value other life-forms, our valuation is biocentric, and ecocentric, when we value ecosystems for their own sake. At least sometimes environmental philosophy, which values non-human nature for its own sake, is called biocentric to distinguish it from an anthropocentric ethics, according to which only human beings are valued for their own sake while all other things are valued for the sake of people.

An anthropocentric thinker would accept naturogenic values from the ontological point of view, if these naturogenic values were well-reasoned. Or some naturocentric thinkers may assume that all values are generated by people, but their ethical argument would give priority to nature. In their mind, nature is more valuable than people. Which comes first in a conflict situation, nature or people? If you wish to be a proper naturocentric thinker, you should defend the thesis that nature is more valuable than people. An anthropocentric thinker can value nature both intrinsically and instrumentally, unless no harm will be caused for people. In the conflict situation, you may defend human value over and above natural value.

All kinds of anthropocentrists attribute more value to human life than animal and plant life. Human life is thought to have a special value, sanctity or dignity, that no other life-forms have. The idea of the intrinsic value of human beings is often argued by philosophical Christianity, in which the primary source of value is God. Only human individuals as unique beings who are made in the image of God, deserve to be valued intrinsically. Human value consciousness has the highest intrinsic value. For instance, Hartman claimed that human beings, as unique centers of consciousness with their evaluating capacities, form intrinsic values, while the deepest source of value is God.[2] Many people believe God is the origin of value. The belief of human culture is the origin of value is not a common view. We also should consider another possibility: animals and nature as the origin of value. In the following, I discuss the difference between ontological and ethical anthropocentrism. Ontological anthropocentrism claims that no naturogenic value exists. A kind of ethical anthropocentrism, ecohumanism, is an environmentally acceptable position in taking the non-human environment into account for the well-being of people. As it is an anthropocentric position, it argues for human priority over the non-human world. A genuine naturocentrist thinker's claim is that the intrinsic value of nature takes priority over human intrinsic value.

1. A Critique of Anthropocentrism

I distinguish strong anthropocentrism from weak anthropocentrism concerning the question of intrinsic value in the following sense. According to strong anthropocentrism, nature is valueless until people ascribe some instrumental value to it. Nature cannot have intrinsic value. Only people have this special kind of value.

Weak anthropocentrism attributes intrinsic value in terms of ends-value to nature but supposes that the intrinsic value of people is prior. Weak anthropocentric positions based on the anthropogenic theory of value ascribe intrinsic value to animals and nature. Such value is ascribed to, as opposed to found in. In both strong and weak anthropocentrism, most significant values are connected with human good and well-being. The question of "what is value?" is understood in terms of what people actually appreciate.

In axiology, the locus of value and the priority of value are two different topics. The thesis of which kinds of entities deserve intrinsic value is a qualitative statement of value-objects. We need ethical criteria or requirements for those who deserve intrinsic value and for those who not. Regan's suggestion is that the subject-of-a-life is the basic requirement for an entity having intrinsic value. Those animals, including human beings, which are subjects-of-their-own-life, have intrinsic value. Taylor goes further. The sufficient requirement to have intrinsic value for an entity would be the ability to have a good-of-its-own. In Taylor's view, each living entity satisfies this requirement, and, thus, is the locus of intrinsic value. The anthropocentric thesis that the intrinsic value of human beings takes priority over the intrinsic value of everything else is a comparative statement of value-objects. It says that people are more valuable than animals or plants. The criterion lies in the rich and complex conscious and experiential states of an entity.

Strong anthropocentrism denies the intrinsic-value view of nature. According to it, only human beings have intrinsic value. The question of the priority of intrinsic value does not arise at all in this context, since human beings are claimed to be the only cases having intrinsic value. Discussion of the intrinsic value of nature is ruled out of existence in strong anthropocentrism. Weak anthropocentrism is a more acceptable approach. In this view, discussion of the intrinsic value of nature is not prohibited. Both animals and nature may be regarded as the holders of intrinsic value, but human values take priority over natural values.

Strong anthropocentrism presupposes that nature is created or that it exists for human purposes. Values are related to human beings in the sense that people create values independent of their environmental context. Strong anthropocentrism supposes that values are individually and socially constructed. No necessary criterion of value is found in physical or natural environments. A basic challenge to these kinds of strong anthropocentric ethics is that anthropocentric ethics has ignored the environmental context of human life. The concept of human good must be understood its social context. A human being is primarily a social animal. A good human life is not possible independent of other human beings. I add to this that human good cannot be understood without deep analysis of human relations to the environment. To ignore the environmental context is to neglect an essential part of human good. The view that nature has nothing to do with human morality ignores the evident truth: human life in all its aspects, including its moral aspects, is tied to its physical and natural environments. Moreover, the states of non-human nature determine the structure of human societies, just as human life determines the states of non-human nature. Nature is not an isolated, a passive object to be perceived or utilized by human culture, but the interactive unity of human life.

Our valuing is not an isolated occurrence in human brains, but it occurs in relation to the valuable objects outside our bodies and brains. The claim that animals and nature are something that we can value solely instrumentally is inconsistent with the empirical reality, in which we appreciate different kinds of objects that may be either instrumentally and intrinsically valuable. In the anthropocentric context, all values are related to people, but not exclusive to them. A human relation to nature is an internal relation. A scientific view of nature, following the scientific revolution of the seventeenth century, is based on the supposition of the possibility of a human external relation to nature. A new ecological view, with its old philosophical roots, as recently described by deep ecologists and some environmental philosophers and environmentalists, is based on a view of interactive human relations with nature.

Strong anthropocentrism is wrong in supposing that nature cannot have intrinsic value for human beings. Two arguments are as follows. First, from the truth that the human being has coerced other species into being instruments and tools for human purposes, it does not follow that they are valueless in themselves. Second, human beings have the ability to value things instrumentally, and intrinsically in the meaning of ends-value. To defend the view that nature cannot have intrinsic value in this anthropogenic sense is difficult if not impossible. By ignoring the significance of animals and nature strong anthropocentrism is not even interested in real human well-being. It may be interested in the human economy and industrial development, but not the well-being of people or societies in their cultural, religious, aesthetic, and environmental aspects. Anthropocentrism which is properly concerned with human well-being cannot ignore the significance of the environment and the non-human world for human well-being. I call those views which emphasize the true well-being of people weak anthropocentrism.

2. Weak Anthropocentrism

Some philosophers have wrongly supposed that anthropocentrism means rejecting all kinds of intrinsic value in nature. Taylor contrasts the human-centered (anthropocentric) to life-centered (biocentric) theories of environmental ethics.[3] According to his biocentrism, living beings have inherent worth, while anthropocentrism rejects this value. Andrew Brennan connects anthropocentrism to the instrumental-value view.[4] We may have a more positive picture of anthropocentrism than those who think that anthropocentrism means rejecting all intrinsic values in nature. We should redefine anthropocentrism so that according to it we can ascribe intrinsic value to non-human beings or things. It is insufficiently concerned with the well-being of the non-human world.

Anthropocentrism means enhancing and fostering humanity before the welfare of other beings. The highest ethical goal is to defend the welfare of people. A. T. Nuyen is a typical anthropocentrist.[5] He is a speciesist but not a human chauvinist. He stresses differences between people and non-human beings, but he does not claim that people are superior to other species. Anthropocentrism is present when we ought to promote and enhance human good and dignity. His view differs from Kantian ethics because for Kant duties to animals are indirect duties toward humanity, while Nuyen's duties are strict duties to animals and nature. But both theories are still anthropocentric types. In both cases, we care for animal suffering only for the sake of people, not for the sake of these animals which are suffering. Cruelties to animals increases cruelties between people in society (as Kant would say). Nuyen separates duties to other human beings from duties to non-human beings and things.

Evidently we enrich our human life by preserving the natural world, as Bryan Norton maintains.[6] Norton denies the intrinsic-value view of nature. Trying to attribute intrinsic values to non-human beings and things is not necessary because we have good anthropocentric grounds for nature preservation since in addition to demand or use-value, nature has transformative value for us. This kind of idea of value can be found in transcendentalism. The presence of human beings in nature can transcend the non-human world by bringing forth its intrinsic value and beauty. Transformative values mean values which educate and transform us. In Norton's example, suppose that children playing in the woods are destroying the eggs of birds they find. An adult explains that eggs are necessary to hatch baby birds and shows the children some young birds. In this situation, the encounter with birds has transformative value. The problem is the presupposition that preserving birds is more worthy than destroying them. Nature preservation elevates and educates human values, a view with which I agree. But some people do not appreciate or respect nature and non-human things. They suppose that these things are for human uses. On his subjective value perspective, Norton has no argument as to why his values are better than another's values. Somebody might think that destroying the eggs of birds educates boys. It transforms them into tough and aggressive men who will do well in the rough and tumble world. If people sentimentally care for the eggs of birds that is needless sentimentality and dangerous for the development of human culture. The idea of transformative value (such values and things elevating other values) is evidently right. This is close to the idea of contributive values, those values that contribute to other values.

Giving value to animals and nature means ascribing or attributing values to them. This must be distinguished from the view that animals and nature generate values in themselves. To say that clean water is valuable means for subjectivistic thinkers that somebody gives, ascribes, or attributes value to clean water. But for objectivistic thinkers, it means that clean water has the value of clean in itself. The problem of strong subjectivism is that if somebody says that she values this water, I can ask of her, "but why do you value this water?" She should refer to qualitative properties of water to answer this. Brennan maintains that non-humans are important to us as such, and not only for our own welfare.[7] For him, ethical extensionism, and even naturocentrism (ethical concern for non-human animals and nature) is thereby intelligible.

Hargrove has defended the weak anthropocentric intrinsic-value position.[8] By the term "weak anthropocentrism" he calls attention to anthropocentric intrinsic-values, noting that not all valuing is instrumental. On this I agree. People can ascribe intrinsic values as well as instrumental values to the non-human nature. Both types of values are related to human beings. A person values non-human nature instrumentally or intrinsically. But I do not agree with Hargrove's suggestion that speaking of non-anthropocentric values is unnecessary since all values are human values. Although our point of view is always the human point of view, it does not follow that all human valuing is anthropocentric. To define anthropocentrism to mean that our point of view is always the human point of view is a tautology and therefore an uninformative definition. There is no contrast class. No human being can possibly have a non-anthropocentric view. Anthropocentrism, in a more helpful sense, means a broad ideology according to which people are seen as basically more significant than other beings on Earth. In this redefinition of anthropocentrism, its basic thesis is that people are more valuable than nature. Animals and nature may matter for you, but they are less significant than other people. Anthropocentrists even can value non-human beings for their own sake and promote their goods as ends in themselves, because to recognize and promote natural goods enriches human life. For instance, John O'Neill's view is a typical anthropocentric view that is appropriately justified, but he attacks naturocentric positions while misunderstanding them.[9] He claims that to defend nature's intrinsic value we should show that such value contributes to the well-being of human agents. This is intended to persuade us to reject the naturocentric position, not being a valid argument against it. Naturocentric positions do not focus on the well-being of people but try to show that we must take into moral account well-being in the non-human world. We must do it regardless of whether it contributes to human well-being. If somebody defends an altruistic position, you cannot argue that the person needs to show that his or her position contributes his or her own well-being. And for the naturocentrist defenders of nature's intrinsic value, you cannot argue that they need to show that such value contributes to the well-being of human agents. This is the task of anthropocentrists. The naturocentrist is not basically interested in the well-being of human agents only, just as the altruist is not basically interested in his or her own well-being. If they were, they would not defend such positions. A theoretically ideal situation is that we might care both about non-human and human well-being, and believe that they are connected. In practical life, such ideals hardly function. Preserving nature means harm for some people.

From the anthropocentric point of view, we have no direct duties to animals. If you are not cruel to a dog, you do so because of humanity. You think that cruelty to animals increases the immoral development of people. The redefined anthropocentrism approves the intrinsic value of animals. It imposes duties for us to take care for animal well-being.

3. Ecohumanism

In the context of environmental issues, the most promising model of anthropocentrism is ecohumanism. In humanism, human beings are the most significant beings on Earth. This gives a false picture of the world in putting human beings into a position they do not occupy. David Ehrenfeld has described this arrogant human attitude to nature.[10] The roots of humanism can be found early in our history. The arrogance is abundantly alive in human practices. We human beings have a unique ability to exceed our human reality: dogs are necessarily dog-centered, but human beings have the ability to know something of what it is like to be a dog. Our ability is limited. We never know what to be a dog as a dog is. All our knowledge is human knowledge of the world, however scientific that knowledge may be. Yet we know that we are not the only inhabitants of Earth. We share it with diversified fauna, flora, and fungi. The conventional estimate is that three to five million species inhabit Earth. A beetle specialist, T. L. Erwin, found that typical tree canopies in tropical forests contain about 1,000 species of beetles. Erwin's estimate of insect species is 30 million in total, so that the total sum of species exceeds three to five million species. Only 1.5 to 1.8 million species have been recorded and given Latin names.[11] We are not alone in the world, although human beings have behaved as if we were.

Some remarks on humanism and anthropocentrism are needed. An anthropocentric attitude to nature is necessary to be a humanist, although this attitude is only part of the ideology of humanism. A humanism can be understood as more broad and complex than anthropocentrism. Two basic historical features of humanism are respect for human beings and respect for human culture. The basic feature of humanism is to defend the good of human beings. The aim of humanism is the *regnum hominis*.[12] My critique of humanism is that it over-stresses the significance of human beings and disregards the significance of the non-human world. Living in the *regnum hominis* seems impossible. Human life requires the *regnum bioticum* (or *biologicum*) in many ways and thus respect for animals and nature.

I am not sure whether humanism or anthropocentrism is really an ideology of people in general. From the historical point of view, humanists are normally men (for instance, John Stuart Mill, Francis Bacon, Karl Marx). It may be correct to speak of androcentricity instead of anthropocentricity, since I wonder whether female thinking is recognized at all in humanism. Although I think that women also subdue the non-human world, men evidently have had more power and influence to do this in a historical and practical sense. Humanistic values are classified as soft values which are opposed to economical and technological hard values. In this classification, ecological values and moral values are stereotypically soft values.

Could humanists have a naturocentric attitude to nature? I argue that they cannot, since humanism means to have an anthropocentric attitude to nature, to be interested in human welfare, the good of people, above the well-being of the non-human world. Humanists, by definition, have to believe that the highest and the most central place on Earth belongs to the human species. Humanists clearly demarcate even between human beings and the highest animals, while naturocentrists see similarities between human beings and all life-forms, all living beings having their own good that should be respected.

Two interesting models of humanism are both called ecological humanism. Skolimowski has developed ecological humanism in *Eco-Philosophy*[13], Brennan in *Thinking about Nature*[14] has defended what he also calls ecological humanism. While the two views are quite different they are similar as paradigm cases of the anthropocentric views. Skolimowski argues that we need a new moral insight: we are a part of nature, and nature was not created for our use. For Skolimowski, nature cannot have intrinsic value; only our human value consciousness has intrinsic value. This is a strongly subjectivistic attitude to nature. Skolimowski defines values such that intrinsic values in nature become impossible: all values exist in human consciousness. All other kinds of values are ruled out *a priori*. The intrinsic value of nature is denied existence.

Skolimowski's view resembles the Kantian view. For Kant, a good will was claimed to be the bearer of an intrinsic value or inner worth.[15] For Skolimowski, the most important intrinsic value is the human attitude of "reverence for life." The other basic intrinsic values for him are responsibility and frugality. Skolimowski is establishing a new ecological axiology.[16] Ecological values (diversity, richness, complexity, eco-justice) are the new values of our times. Skolimowski claims that some human values and attitudes are better than others. Ecologically-based attitudes are better than more traditional attitudes which ignore ecological values.

The basic difference between anthropocentrism and naturocentrism is that naturocentrism rejects the view that human well-being always takes precedence over animals and nature. We can, and from the naturocentric position we should, take non-human nature into our moral scope whether it contributes to human good or not. From the ethical extensionist point of view, animals and nature are not made for or by humanity. People have no human right to use these animals and natural areas as they please. Animals and plants life for animals and plants is the only life they have, just as human life is for human beings. If we respect this point of view for human life, we should respect other life-forms. I support this ecohumanism respect for animals (including human animals) and nature. The idea of humanity is formulated in the Kantian moral imperative: act in such a way that you always treat humanity, whether in your own person or in the person of any other, never simply as a means, but always at the same time as an end. We can translate this into the ethical extensionist or naturocentric language as follows:

> Act in such a way that you always treat nature, whether in your own life or in the life of any other, never simply as a means but always at the same time as an end.

This could be called the ecological imperative. If we connect these two imperatives, the conclusion is a form of ecohumanism. We emphasize values of humanity and at the same time ecological values. The paradox of humanism is that while it has created our present culture on the assumptions of human power, many humanists may feel a closeness or kinship with nature.[17] In ecohumanism, the paradox of humanism disappears, since both sides have been taken into account. The idea of humanism respects the good of human beings, while the idea of ethical extensionism respects the good of nature. While strong anthropocentrism stresses differences between human beings and nature, ecohumanism connects the two aspects of humanity and ecological values. It requires an attitude of respect for animals and nature, because it requires ascribing anthropogenic intrinsic value to them.

4. The Problem of Priorities: Are People Always Superior to Nature?

The question of priorities remains a problem of ecohumanism. As an anthropocentric view, ecohumanism should consider the intrinsic value of people prior to the intrinsic value of nature. In a genuine naturocentric view the non-human world should win the decision in any conflict. On the naturocentric perspective, we are defending the primacy of the existence of nature over human goodness. The hierarchy of values that puts ourselves on top, ascribes all "excellences" to being human. Can we rationally justify human excellences? How far is "human intrinsic value" really distinctive of human value, which cannot be ascribed to any non-humans?

Kant defended the special dignity of human beings, as opposed to the relative worth of anything else. People have an inner worth, only people are morally good and rational beings. Christine M. Korsgaard calls this idea the Formula of Humanity as an End in Itself.[18] What is the justification for inner worth as an intrinsic value of human beings? Why should our power to justify our ends be something that has intrinsic value? Kant has already presupposed that humanity is the supreme value. He gives no argument why only humanity, the uniquely human dimension of myself, should be the highest value in my life. Why is it not just as important to recognize our relations to the non-human world within which we live our every-day lives? If I regard myself as a conscious and a living entity and thus as having inner worth, then the Kantian doctrine can be formulated that we should treat all conscious and living beings as having the same inner worth in virtue of their consciousness and life. To assume that exclusively human beings have intrinsic value and that this is because they are human beings, that being a human being is automatically more valuable than being some other being, is speciesism that we should reject. Special human values occur in the world, but human value is not the lonely intrinsic value present on the earth, and not the greatest intrinsic value.

In this book I have argued that nature has intrinsic value in many different senses. To say that "an animal has intrinsic value" is a qualitative type of statement in axiology. How can we then answer the comparative question of the hierarchy of naturocentric values: has an animal more intrinsic value than a plant? A conventional ethical extensionist hierarchy would be that zoocentric values have higher moral standing than biocentric values, and biocentric value is morally superior to ecocentric value. First we have to take the level of animal life into moral account; second, the level of organic life; and third; the level of ecosystems.

I prefer the opposite hierarchy, in which the ecogenic level is the most important because it is the ground of all other levels: it generates the levels of life-value and conscious value. The land, ecosystems, and the earth have a systemic value. Is the ecogenic value level the most significant level of the value of nature? According to this argument, then, biocentric value would be superior to zoocentric value. Life is the ground of consciousness. It generates the level of the value of consciousness. The level of organic life could have more value than the level of sentient life. But do we really want to argue that plants have more intrinsic value than animals? We don't need to. We can instead say that plants and animals both have organic value and that most animals have in addition the value of consciousness.

The intrinsic value of consciousness is a universal intrinsic value which human beings share with other animals. The intrinsic value of livingness is another universal intrinsic value which we share with all life forms. A third universal intrinsic value is nature as a whole of which we are part. If consciousness is an intrinsic value, then *Homo sapiens* is an instance of the group of conscious beings which carry this value. That animals acquire their value from the intrinsic value of human consciousness is not true as is the contrary: human beings get their intrinsic value from the universal intrinsic value of consciousness, livingness, and wholeness. We get the value of consciousness from the value level of sentient beings. We get our consciousness value from being animals.

Weak anthropocentrism is based on the view that people have superior value to other beings, while in strong anthropocentrism humanity is the only end in itself. There is no viewpoint of non-human beings that should be taken into moral account. In my naturocentric argument, human beings derive their intrinsic value from being a part of the universal intrinsic value of consciousness, of livingness, and of ecosystems. In this sense, people have intrinsic value at three levels. This is a qualitative statement of what kinds of intrinsic value human beings have. The comparative argument is that nature as a whole has more value than livingness, and livingness has more value than consciousness. The basic difference between anthropocentrism and naturocentrism is that naturocentrism rejects the view that human well-being always takes precedence over animals and nature. We can, and from the naturocentric position we should, take non-human nature into our moral scope whether it contributes to human good or not. From the naturocentric point of view, animals and nature are not made for or by people. People have no human right to use them as they please. The distinction between anthropocentrism and naturocentrism remains evident in that anthropocentric theories stress the importance of human beings, human values, and related things, while naturocentric theories focus on animals (other than human beings), life, and ecosystems, and they ask: What is the intrinsic value of nature?

Nine

THE RIGHTS OF ANIMALS AND NATURE

Some ethicists have denied the rights of nature. Intrinsic value with the attitude of respect includes the idea that those beings which have intrinsic value are worthy of respectful treatment. Using the language of rights would not carry any new concepts or ethically significant ideas.[1] Another possibility is the trend that moves from natural rights to the rights of nature.[2] I argue in this chapter that moral rights have a key role in practical and.political morality. Giving rights to animals and nature is supported by practical arguments. Their legal rights would be an effective tool to preserve them.

The conceptual link between values and rights appears clearly in such expressions as "value in their own right" and "valuable in its own right." The expression "valuable in its own right" refers explicitly to the class of beings which are of intrinsic value. Bernard Gendron perceives a strong conceptual relationship between intrinsic value and rights: anything that can be intrinsically valued has moral standing, and can have rights. He argues that values and rights exist only in our human verbal life, but have no ontological nature. Intrinsic value refers to something which is valued as an end-in-itself, in opposition to things which are valued instrumentally: People value some non-human things intrinsically, other non-human things instrumentally, and some non-human things in both ways. To say that they have intrinsic value is to say that we have good reasons for valuing them as end-in-themselves.[3] Gendron assumes that those who deny nature's intrinsic moral value believe that intrinsic value is a property that we discover in things.[4] For Gendron, intrinsic value is a verbal rather than a substantial phenomenon. In this anthropogenic-value view, instead of discovering values from nature we ascribe them to nature. According to the naturogenic-value view, values are ontologically intrinsic to animals, plants, and ecosystems. Many people have a sense of values which are intrinsically in nature. We may defend those values which exist on the ontological level. Sentient animals have the right to respectful treatment because of the kind of value they have. The moral rights in this sense would be called "intrinsic moral rights," meaning those rights which something has on the ontological level. Two forms of value are the squirrel as a value for human beings, and the value of the felt experience of a squirrel as what it is to be the squirrel, that is, the individual center of consciousness.[5]

The term "rights" refers to the idea that the holders of rights are of intrinsic value and have to be respected. The talk of rights would be senseless if nothing followed from rights. Rights correlate with duties. If I have a right to life, then others have a duty not to kill me; if I have a right to publish a book, then others have a duty not to prevent me publishing the book (although nobody may have the duty to publish my book). If something has intrinsic value, that is a reason for a duty. But if other reasons clash, there may be no duty (or right either). In the following, I present an example of the argument that our duties to the non-human world can express that it has intrinsic value. Our duties do not generate the intrinsic value of nature. The language of duties makes it easier for us to perceive intrinsic values in nature. The basic premise is that those to whom we owe duties are not mere instruments for us, but have intrinsic value which we have to respect. We can assume that we have no duties toward beings or entities which have no intrinsic value. And those beings to whom we owe duties have some intrinsic value; they cannot have only instrumental value for us. Toward instruments, we need not owe any duties, since we are allowed to use them for our purposes. Thus, if we have some duties to animals or other non-human beings, then they have some intrinsic value and cannot be merely a means for us. The argument is as follows:

(1) All those to whom we owe duties are not only instruments for us but have their own value, that is, are of intrinsic value.
(2) We have duties to all those to whom we can cause suffering and pain by our action.
(3) Our action can cause suffering and pain to sentient beings; therefore we have duties to sentient beings.

Therefore,

(4) Sentient beings are not only instruments for us but have their own value, that is, are of intrinsic value.

This type of argument would also express the intrinsic value of trees and plants, if the second premise were to be, "we have direct duties toward all which we can harm or benefit," and if we presume that we can harm or benefit any living being.

Intrinsic values, rights, and duties are entangled. Attributing these concepts to non-human beings depends on whether our world-view is anthropocentric or naturocentric. Anthropocentric views are more unwilling to attribute these concepts to non-humans than are naturocentric views.

1. The Rights of Human Beings

In Kantian ethics, everybody has the right to respectful treatment and nobody is allowed to be treated merely as a means for others' purposes. All human beings have intrinsic value and certain rights which have to be respected. In anthropocentric views, moral rights are restricted to human beings. The concept of rights do not include animals or other non-human beings. An argument is that rights are indispensable to defend the extraordinarily high value of human beings.[6] Rights must be restricted to human beings since people have greater worth than any other beings. Plants and animals do not have the same degree of value. Rights can be overridden only by more primal rights. Since human beings have the right to kill animals and plants quite freely, animals and plants can have no right to life. Rights secure human individuals. Trees are raw-material for books, and lab-animals are raw-material for experiments. If these plants and animals had the right to a life, they could not be killed for human purposes.

John Passmore has argued that the rights of non-human beings are nonsense because non-human beings do not belong to the same communities as people and that they do not satisfy the criterion of reciprocity. They do not share interests and obligations with people.[7] A third reason sometimes offered is that rights are inherently related to human language. Non-human beings cannot claim or justify any rights because they have no language.[8] Three counter-arguments for these claims follow.

1. From an ecological point of view, people, animals, and plants do have some mutual biological interests. For example, they need a safe environment, water, and nutrients. A basic idea of environmental ethics is that human beings share their living environments with other beings. We live both in human communities and in biocommunities. Human welfare is integrated with the welfare of non-human beings. Human and non-human interests are not always conflicting. For example, we need bacteria for our human welfare, and these small organisms, like many others, have an important function in the life-support systems of this planet.

2. Not all human beings, such as, small children, criminals, and sick persons, satisfy the criterion of reciprocity. Nevertheless, we treat them morally, and the same may hold true regarding non-human beings. Animals who live with us have some obligations toward us (although these may not be moral duties). Dogs and cats are not allowed to bite us or to stray too far away from home, cows are obliged to give us their milk, and so on. We have a strong moral and cultural relationship with many animals, although in Western history this relationship has been mostly human domination of them.

3. We should also refute the point which says that only those beings who can claim and justify their rights can have rights. The cases of the rights of children and other minority groups indicates that their rights are not necessarily related to their linguistic ability or their intelligence. Some non-persons, like corporations and ships, have rights. Having rights does not necessarily require linguistic abilities.

The relationship between rights and intrinsic value is less studied and more controversial than the relation of intrinsic value to "good" in philosophical literature, as well as in environmental philosophy. Although many environmental philosophers understand the biocentric notions of rights, the term "rights" is normally perceived to be more individualistic and anthropocentric than the terms "intrinsic value" and "good." The concept of rights is a distinctive concept of Western ethics. Although philosophers have discussed rights for years, rights still have a controversial role in modern philosophy, even apart from their use in environmental philosophy. Some philosophers have argued that moral rights bring nothing new or worthwhile to ethics. To know what is right or wrong, good, or bad, is enough. Philosophy has provided a valuable critique of the moral-rights arguments. The most likely candidate to replace the discussion of moral rights is said to be the concept of intrinsic value.[9]

Rights might be necessary to protect individuals in the situation in which fundamental conflicts between one or more individuals and society exist. The basic needs of individuals are not allowed to be subjugated to the public good. This claim is independent of how great the societal utilities are that might be gained if the individual member were subjugated. Our moral relationship with sentient animals cannot be sound without accepting their moral and legal rights.[10] I center next on rights for animals.

2. The Rights of Animals

Animal rights and other philosophical questions about human relations to animals have been extensively analyzed in recent years. An invaluable source is Charles R. Magel's *Keyguide to Information Sources in Animal Rights*.[11] Other important sources are the special numbers on "Animal Rights" in *Inquiry*, and "Animal Rights" in *The Monist*.[12] Several philosophers have defended animals against modern factory-farms and other animal production methods or against scientific experiments which instrumentalize animals for human usage while ignoring the rights of animals to painless, free, and enjoyable life.[13] Some philosophers have written against animals' rights.[14] Typically, even they do not want to defend modern factory-farm methods that cause suffering to animals. My suggestion in this chapter is that animal rights should be understood as a practical method for taking animals into moral and legal account.

The old idea of natural rights is the legal theory according to which human beings have rights by nature. In 1791, the French Constitution espoused the idea of the natural and inalienable rights of man: liberty, prosperity, security, and the resistance to oppression.[15] Setting aside here many philosophical problems of the natural rights theories I think that this old philosophical idea of natural rights or natural law would be a fruitful model for animal rights in that an animal's right to a painless life arises from the fact that animals are sentient beings by their nature.[16] Then "natural animal rights" accord rights to animals as sentient beings. Human rights are more complex. The model of basic animal rights could be a simple model based on sentient animal nature.

Legal animal rights would be the most effective tool for taking the intrinsic value of animals into moral account. Animal activists demands that the rights of sentient beings should be better recognized in the law. The core of such rights should be the principle of equality. How much weight should animal interests deserve in the law? Animal rights activists do not claim that animals should have human rights. Their claim is that animals should have animal rights. The most important animal right would be the right to psychical, physical, and social well-being. This includes the right to a normal animal life of its own kind, including

the right to appropriate liberty
the right to individuality and to social relationships
the right to be free of pain
the right to life.

The right to normal animal life means the right to live as the animals that they are. The real pig's life is different from the real horse's life. Stephen Clark has proposed the valuable idea of "self-owners," animals as sentient beings having their own self-significant life.[17]

Many philosophers, among others, have argued that animals cannot have rights, because legal rights are made by humans beings, as the products of human history, culture, and politics. If we human beings want, we can extend legal rights to animals as well as to other non-human objects. It is a question of a human decision, though strong human interest in not giving rights to animals and nature do exist. This interest is a human interest as well as the interest of the animal rights movement for in giving rights to animals. People should have legal rights to protect animals. This human right could be asserted best by giving legal rights to animals, so animal-rights activists would have an effective tool to defend animals against cruelty and inhumane treatment. We have strong property rights in our legislation. The tragedy is that sentient, living creatures have no rights as the owners of their own lives, for they are treated as the property of people.

The significance of Darwin's theory of evolution for animal rights is that it challenges the view of the uniqueness of the human being.[18] The human being is an animal, which, like the wolf, is descended from other animals. A Declaration on Great Apes demands equal rights for all great apes: the right to life, the protection of individual liberty, and the prohibition of torture.[19] We should agree with this declaration as one step toward animal rights. I find mistaken to suppose that only those beings closely related to human beings such as the great apes should have equal rights with human beings. The declaration demands the extension of the community of equals to include human beings, chimpanzees, gorillas, and orangutans. In my view, we should not give them rights insofar as they are more or less like us, but because of what they are in themselves, sentient beings who value their own lives. Why should only the great apes form the larger moral community with moral and legal rights? To include the great apes within the moral community is an extension of morality but a minimal one from the perspective of other animals and plants. From a zoocentric perspective, I find difficult to assume that we should act morally on behalf of great apes only, not for other animals. All sentient beings should be included in these "great apes principles." All sentient animals should have the right to life, the protection of individual liberty, and the prohibition of torture. In practice, we should confirm the category of sentient beings and bestow legal rights on them. In our modern laws, animals are categorized as property or raw-material, which is evidently inadequate in the light of the understanding of animals as sentient beings.

Ascribing intrinsic value and rights to non-human beings does not mean that they should have all the rights people have. Human rights might be stronger than the rights of animals, plants, or ecosystems, although we would have *prima facie* duties toward them. Our duties to promote their intrinsic value are not absolute but can be overridden.[20]

Those who think that nature has only instrumental value seem to hold that nature is rightless. For example, Passmore has called mystical rubbish the supposition that destroying non-human beings is intrinsically wrong. Those who think that nature is intrinsically valuable are more sympathetic to the claim that nature has rights. We may detect a trend moving from natural rights to the rights of nature. The term "rights" implies that the holders of rights are of intrinsic value and have to be respected. The domain of the holders of intrinsic value and that of the holders of rights is not necessary the same. Ascribing intrinsic value and not moral rights to some things may be possible. Speaking of the rights of nature is not necessary, although we can ascribe rights to sentient animals. We are able to realize how valuable nature is without postulating rights as part of it. Sentient individuals desire rights, and other non-human beings intrinsic value.

The basic question is what kind of entities can be the holders of intrinsic value and rights, that is, to whom we can attribute the terms "intrinsic value" and "rights"? Can we classify entities in accordance with their ontological properties into those which have intrinsic value and rights and those which have not? Joel Feinberg argues that only those beings capable of being represented (that is, having interests, which others can take into account), can have rights. To represent a being that has no interest is impossible; thus, a right-holder must have interests. Feinberg defines a being without interests as "a being that is incapable of being harmed or benefitted, having no good or 'sake' of its own."[21] Then, only the sorts of beings which have interests can be represented and can have rights, since "a being without interests has no 'behalf' to act in, and no 'sake' to act for."[22] Four criteria for rights are presented to follows:

(1) A right-holder is an identifiable individual,
(2) that has interests,

(3) although interests are not sufficient to entail corresponding rights, but
(4) if a right follows from an interest, this depends on the characteristics of a right-holder rather than on the characteristics of others.[23]

Criteria (2) and (3) mean that interests are necessary but not sufficient for having rights. Criterion (4) means that the right of a horse does not depend on whatever interests the owner of the horse, for example, may have. Sentient beings evidently have interests. Rights may be restricted to sentient beings that have interests. Their ability to feel pain obliges us more than the intrinsic value of plants.

One purpose of moral rights is to protect their bearer from harm: X has a right to Y only if the deprivation of Y would (other things being equal) harm X.[24] To be harmed presupposes consciousness; accordingly, rights can be attributed to sentient beings. This suggests that values cannot exist outside the world of experiences. A wilderness area as a whole is rightless since it is not a being with consciousness. But we should doubt whether wilderness is valueless.

The above arguments suggest that only individual, sentient beings can have rights. The relationship between rights and interests is justifiable. Those who have interests can have rights, because they have morally legitimate claims about how they are treated by others. If an entity has no interest in how it should be treated by others, it cannot have any rights to be treated in a human way. We may justifiably agree with this, although showing that sentient beings have intrinsic values and rights does not show that non-sentient beings do not have them.

Ethics based on the rights of sentient beings cannot be a sufficient ground for environmental preservation. Species and ecosystems need to be respected and protected. Individualistic morality is inconsistent with the demands of nature conservationists. This leads to the separation in discussion between animal rights and environmental preservation. Nature-conservation philosophy has to be holistic in defending species and ecosystems. The distinction between animal rights and environmental preservation exists. Yet these issues should not be sharply distinguished; instead, we should treat them as complementary issues.

3. Rights as Practical Tools in Environmental Policy

Gendron's argument is an original view of rights. Granting rights to biotic systems indicates that we can know what their interests are. Rights come first, and interests follow from them. After deciding that a lake has rights, we then know that its interest is what it is to be as a lake. Evidently, its interest is not to be polluted or destroyed. Gendron's argument for interests and intrinsic value is more radical than the views which extend interests to all living beings. He suggests that in addition to individual living organisms, ecosystems, such as lakes, mountains, and forests, have interests, and that we human beings can know what their interests are. Gendron's argument is more a practical than an ontological argument.

If natural objects have legal rights, then the whole legal system will become more protective of the environment than the system in which only human beings have rights. The legal rights of nature will facilitate taking into account all the costs of human activities. Instrumental reasons could then generate belief in the intrinsic value and moral rights of nature. Nature can be valued as an end-in-itself when we have strong instrumental reasons for this kind of valuation, claims Gendron.[25] Valuing nature intrinsically would serve as the basis for preserving nature for ourselves and future generations. Sometimes we do not perceive, before it is too late, that nature's values are threatened by human activities. Gendron highlights that valuing the environment intrinsically provides extra incentive to save it for future generations, since the rights of present biotic systems are more palpable than the rights of distant future human generations. We have reasons to value nature intrinsically on instrumental grounds, until this kind of valuation serves as more effective protection for these natural areas which have values for us. If we confirm that nature has intrinsic value, then we protect it better with fortunate results for ourselves.

Rights are a valuable tool for solving environmental problems. The environmental rights of human beings must be defended: People should have a right to a livable environment because this supports realization of other human rights. The environmental crisis challenges views about what our basic rights are, and it may restrict some of our human rights.[26] The practical argument for the intrinsic value and rights of nature is as follows: (1) we have strong instrumental and moral grounds for preserving natural areas; (2) the rights of nature are effective instruments in preserving natural areas; (3) thus, to ascribe rights to nature is reasonable. The term "rights" is ascribed to something which ought to be preserved or protected. In this sense, I could claim that, the Grand Canyon should have rights.

About twenty years ago, speaking of the rights of non-human beings was interpreted as a sentimental and mystical rubbish.[27] The trend toward the legal rights of non-human beings begins with Christopher Stone, a professor of law in California. Thinking in terms of the rights of nature would help us understand our fundamental relationship with nature.[28]

4. Anthropocentric and Naturocentric Rights

Even if we assume that values and rights exist exclusively in our human culture, in addition to the members of our species we can ascribe values and rights to different kinds of living and non-living beings. Anthropocentrists can ascribe intrinsic value to non-human beings, and in the same sense, they can ascribe rights to them:

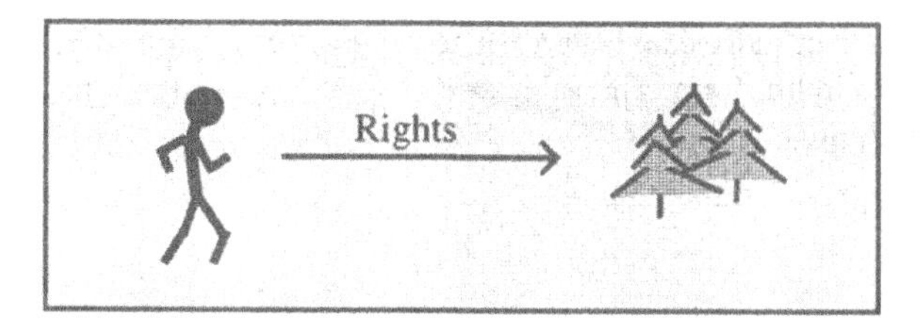

On this anthropocentric view, rights need not be understood in their ontological sense; they need have no ontological nature. The rights of non-human beings are thus possible even if we think that the anthropocentric context is the only possible one. The language of rights is a game we play to get nature conserved for ourselves. A problem is that we can choose some other game, if we wish. If the Grand Canyon has no intrinsic rights, but has only some ascribed or attributed rights, how do we choose what kinds of rights to ascribe to it?

The rights of nature cannot be easily established because two different world-views, anthropocentric and naturocentric, clash. The extension of moral rights from human beings to other beings requires a new, naturocentric world-view: we should recognize our moral relationships with non-human beings.

Naturocentric theories challenge the view that the non-human world is made for limitless human purposes. William T. Blackstone[29] has argued for the right of people to a liveable environment, and Christopher Stone[30] has defended the right of nature to its own existence. Both argue that rights are the most effective tools in the solution of environmental problems. Arne Næss has defended the right of every living being to live and blossom.[31] According to Blackstone, we should understand non-human beings as what they are and how they relate to other things, not only how they function and can be manipulated for human purposes.[32] How we understand the term "rights" is a moral question, and a substantial matter of seeing the world.

Eugene C. Hargrove has predicted that the discussion of the intrinsic value of nature will replace the discussion of rights.[33] I think that the discussion of environmental and animal rights is beginning, not past. The moral and legal rights of animals is a crucial issue in the field of animal welfare. The basic future topic regarding the rights of non-human beings will be the distinctions between anthropocentric and naturocentric rights, and between anthropogenic and naturogenic rights.

New attitudes require changes in how we treat animals and the non-human world and in how we see the world. Real non-anthropocentric theories change the meanings of words. The term "rights" can refer to the intrinsic rights of animals, of trees and plants, and even of mountains, and other ecosystems. Several kinds of intrinsic value exist. The value of sentience requires that sentient beings have the right to live without suffering caused by human activities. The value of individual living beings requires that individuals have the right to respectful treatment in accordance with their individual features. The value of life requires the right to live and flourish in each life-form. The value of ecosystems necessitates their right not be polluted by human activities.

NOTES

CHAPTER ONE

1. Cf. Eric Katz, "Searching for Intrinsic Value," *Environmental Ethics*, 9:3 (1987), p. 235.

2. Anthony Weston, "Beyond Intrinsic Value: Pragmatism in Environmental Ethics," *Environmental Ethics*, 7:4 (1985), p. 321.

3. Klaus Michael Meyer-Abich, *Revolution for Nature: From the Environment to the Connatural World*, trans. by Matthew Armstrong (Cambridge, England: The White Horse Press, 1993), p. 28. [*Aufstand für die Natur, Von der Umwelt zur Mitwelt*, 1990]

4. Bryan Norton, *Why Preserve Natural Variety?* (Princeton: Princeton University Press, 1987).

5. Herbert Marcuse, *One-Dimensional Man* (London: Sphere Books, 1970) [1964].

6. Jacques Ellul, *The Technological Society*, trans. by John Wilkinson (New York: Vintage Books, 1964).

7. Lewis Mumford, *The Myth of the Machine: Technics and Human Development* (New York: Harcourt Brace Jovanovich, 1966).

8. Theodor Adorno and Max Horkheimer, *Dialectic of Enlightenment* (London: Verso Editions, 1979) [*Dialektik der Aufklärung*, 1944].

9. E. F. Schumacher, *Small Is Beautiful: Economics as if People Mattered* (New York: Harper and Row, 1973).

10. Carolyn Merchant, *The Death of Nature: Women, Ecology, and the Scientific Revolution* (London: Harper and Row, 1990) [1980].

11. See Yrjö Haila and Richard Levins, *Humanity and Nature: Ecology, Science, and Society* (London: Pluto Press, 1992), p. 11.

12. See Paul W. Taylor, ed., *Problems of Moral Philosophy: An Introduction to Ethics* (Belmont, Cal.: Dickenson, 1967), p. 410 and Paul W. Taylor, *Normative Discourse* (Englewood Cliffs, N.J.: Prentice-Hall, 1961), p. 26.

13. Cf. C. A. Baylis, "Grading, Values, and Choice," *Mind*, 67: 268 (1958).

14. Taylor, *Problems of Moral Philosophy*, p. 411.

15. Cf. Risieri Frondizi, *What Is Value? An Indroduction to Axiology* (LaSalle, Ill.: Open Court, 1971), p. ix.

16. See Ronald N. Giere, *Understanding Scientific Reasoning* (Orlando: Holt, Rinehart, and Winston, Dryden Press, 1991), pp. 282-283; J. O. Urmson, "On Grading;" *Mind*, 59: 234 (April 1950); and Baylis, "Grading, Values, and Choice."

17. Gilbert H. Harman, "Toward a Theory of Intrinsic Value," *The Journal of Philosophy*, 64 (1967).

18. Charles L. Stevenson, *Ethics and Language* (New Haven: Yale University Press, 1947) [1945], p. 177.

19. See J. Baird Callicott, "On the Intrinsic Value of Nonhuman Species," in Bryan Norton, ed., *The Preservation of Species: The Value of Biological Diversity* (Princeton: Princeton University Press, 1986), p. 143. See also J. Baird Callicott,

In Defense of the Land Ethic: Essays in Environmental Philosophy (Albany: State University of New York Press, 1989), pp. 133, 160-162.

20. Robert Elliot, "Intrinsic Value, Environmental Obligations, and Naturalness," *The Monist*, 75:2 (1992), pp. 142-144.

21. G. E. Moore, *Philosophical Studies* (London: Routledge and Kegan Paul, 1951), p. 260.

22. G. H. von Wright, *The Logic of Preference: An Essay* (Edinburgh: Edinburgh University Press, 1963), p. 7.

23. C. A. Hooker, "Responsibility, Ethics, and Nature," *The Environment in Question*, eds. David E. Cooper and Joy A. Palmer (London: Routledge, 1992).

24. Kenneth E. Goodpaster, "On Being Morally Considerable," *Ethics and the Environment*, eds. Donald Scherer and Thomas Attig (Englewood Cliffs, N.J.: Prentice-Hall, 1983), p. 33.

CHAPTER TWO

1. Jorge J. E. Gracia, "Evil and the Transcendentality of Goodness: Suárez's Solution to the Problem of Positive Evils," *Being and Goodness: The Concept of the Good in Metaphysics and Philosophical Theology*, ed. Scott MacDonald (Ithaca, N.Y.: Cornell University Press, 1991), p. 169.

2. *Ibid.*

3. G. H. von Wright, *The Varieties of Goodness* (London: Routledge and Kegan Paul, 1963), p. 50.

4. See Judith Scoville, "Value Theory and Ecology in Environmental Ethics," *Environmental Ethics*, 17:2 (1995), p. 133.

CHAPTER THREE

1. C. I. Lewis, *An Analysis of Knowledge and Valuation* (LaSalle, Ill.: Open Court, 1946), p. 405.

2. C. A. Baylis, "Grading, Values, and Choice," *Mind*, 67:268 (1958), p. 491.

3. *Ibid.*, p. 490.

4. Michael Zimmerman, "On the Intrinsic Value of States of Pleasure," *Philosophy and Phenomenological Research*, 41 (September 1980-June 1981).

5. Archie J. Bahm, *Axiology: The Science of Values* (Amsterdam and Atlanta: Rodopi, Value Inquiry Book Series, 1993), pp. 42-57.

6. Eero Paloheimo, *Maan tie* (Helsinki: WSOY, 1989).

7. Gilbert H. Harman, "Toward a Theory of Intrinsic Value," *The Journal of Philosophy*, 64 (1967).

8. Rem B. Edwards, "Universals, Individuals, and Intrinsic Goods," in Rem B. Edwards and John W. Davis, eds., *Forms of Value and Valuation: Theory and Applications* (London: University Press of America, 1991), p. 81.

9. Immanuel Kant, "Duties to Animals," *Animal Rights and Human Obligations*, eds. Tom Regan and Peter Singer (Englewood Cliffs, N. J.: Prentice-Hall, 1976), p. 122.

10. Thomas Aquinas, "Differences between Rational and Other Creatures," *Animal Rights and Human Obligations*, eds. Regan and Singer, p. 56.

11. Immanuel Kant, *Ethical Philosophy*, The Complete Texts of *Grounding for the Metaphysics of Morals* [1785] and *Metaphysical Principles of Virtue* [1797], trans. James W. Ellington (Indianapolis, Ind.: Hackett, 1983), pp. 97-98.

12. Paul W. Taylor, *Respect for Nature: A Theory of Environmental Ethics* (Princeton, N. J.: Princeton University Press, 1986).

13. *Ibid.*

14. See John Lachs, "Is Everything Intrinsically Good?", *The Journal of Speculative Philosophy*, 5: 1 (1991); and Erazim Kohák, "Why Is There Something Good, Not Simply Something?", *The Journal of Speculative Philosophy*, 5:1 (1991).

15. R. B. Perry, *General Theory of Value: Its Meaning and Basic Principles Construed in Terms of Interest* (London: Longmans and Green, 1926), p. 45.

16. Robert E. Carter, "The Norm of Intrinsic Valuation," in Edwards and Davis, eds., *Forms of Value and Valuation: Theory and Applications.*

17. Aldo Leopold, *A Sand County Almanac and Sketches Here and There* (Oxford: Oxford University Press, 1968), p. viii.

18. Henryk Skolimowski, *Living Philosophy: Eco-Philosophy as a Tree of Life* (London: Arkana, 1992).

19. Tom Regan, "The Nature and Possibility of an Environmental Ethic," *Environmental Ethics*, 3:1 (1981).

20. R. F. Dubois, *Loup, Mon Ami. Wolf, My Friend* (Outrelouxhe, Belgium: Wolves and Wildlife Society, 1994).

21. Peter Singer, *Animal Liberation: Towards an End to Man's Inhumanity to Animals* (London: Granada, 1977).

22. Arthur O. Lovejoy, *The Great Chain of Being: A Study of the History of an Idea* (London: Harvard University of Press, 1964).

CHAPTER FOUR

1. S. F. Sapontzis, *Moral, Reason, and Animals* (Philadelphia: Temple University Press, 1987), and "Animal Rights and Biomedical Research," *The Journal of Value Inquiry*, 26:1 (1992).

2. Marian Stamp Dawkins, "The Scientific Basis for Assessing Suffering in Animals," in Singer, ed. *In Defence of Animals*, and "From an Animal's Point of View: Motivation, Fitness, and Animal Welfare," *Behavioral and Brain Sciences*, 13 (1990).

3. J. S. Kennedy, *The New Anthropomorphism* (Cambridge, England: Cambridge University Press, 1992).

4. M. F. Stewart, "Teaching of Animal Welfare to Veterinary Students," *The Status of Animals: Ethics, Education, and Welfare*, eds. David Paterson and Mary Palmer (Wallingford, Pa.: Humane Education Foundation, 1989).

5. Michael W. Fox, *Inhumane Society: The American Way of Exploiting Animals* (New York: St. Martin's Press, 1990).

6. Arthur L. Caplan, H. Tristram Engelhardt, Jr., and James J. McCartney, eds. *Concepts of Health and Disease: Interdisciplinary Perspectives* (London: Addison-Wesley, 1981).

7. Leena Vilkka, "On Animal Consciousness," in Paavo Pylkkänen and Pauli Pylkkö, *New Directions in Cognitive Science* (Helsinki: Finnish Artificial Intelligence Society, 1995).

8. Thomas Nagel, "What Is It Like to Be a Bat," *Philosophical Review*, 83 (October 1974), p. 436.

9. Dawkins, "From an Animal's Point of View," p. 2.

10. T. L. S. Sprigge, "Non-Human Rights: An Idealist Perspective," *Inquiry*, 27 (1984).

11. Dawkins, "The Scientific Basis for Assessing Suffering in Animals," p. 40.

12. P. R. Wiepkema and J. M. Koolhaas, "The Emotional Brain," *Animal Welfare*, 1 (1992).

13. Françoise Wemelsfelder, "Animal Boredom: Do Animals Miss What They've Never Known," *Euroniche Conference Proceedings: Animal Use in Education*, eds. Bryony Close, Francine Dolins, and Georgia Mason (London: Humane Education Centre, 1989); Dawkins, "From an Animal's Point of View."

14. Bernard Rollin, "The Teaching of Responsibility," *The Hume Memorial Lecture: 10th November 1983 at King's College* (University of London, UFAW, 1983).

15. Michael Leahy, *Against Liberation: Putting Animals in Perspective* (London: Routledge, 1991), p. 253.

16. James Rachels, *Created from Animals: The Moral Implications of Darwinism* (Oxford: Oxford University Press, 1991).

17. Peter Singer, *Animal Liberation: Towards an End to Man's Inhumanity to Animals* (London: Granada, 1977); Peter Singer, *In Defence of Animals* (Oxford: Basil Blackwell, 1986); Peter Singer, *Practical Ethics* (Cambridge, England: Cambridge University Press, 1986).

18. Singer, *Animal Liberation: Towards an End to Man's Inhumanity to Animals.*

19. Mary Anne Warren, "The Rights of the Nonhuman World," *Environmental Philosophy: A Collection of Readings*, eds. Robert Elliot and Arran Gare (The Open University of Press, Milton Keynes, 1983).

20. See R. G. Frey, *Interests and Rights: The Case Against Animals* (Oxford: Clarendon Press, 1980); Leahy, *Against Liberation*; Peter Carruthers, *The Animals Issue: Moral Theory in Practice* (Cambridge, England: Cambridge University Press, 1992); Kennedy, *The New Anthropomorphism.*

21. R. B. Perry, *General Theory of Value: Its Meaning and Basic Principles Construed in Terms of Interest* (London: Longmans and Green, 1926), and Paul Ziff, *Semantic Analysis* (New York: Cornell University Press, 1960).

22. Tom Regan, "The Nature and Possibility of an Environmental Ethic," *Environmental Ethics*, 3:1 (1981).

23. Tom Regan, *The Case for Animal Rights* (London: Routledge, 1988).

24. Albert Schweitzer, *Elämän kunnioitus*, trans. by Juho Tervonen (Helsinki: WSOY, 1966), [*Denken und Tat*]; Albert Schweitzer, "The Ethics of Reverence for Life," *Animal Rights and Human Obligations*, eds. Tom Regan and Peter Singer (Englewood Cliffs, N.J.: Prentice-Hall, 1976).

25. Regan, *The Case for Animal Rights.*

26. *Ibid.*, p. 236.

27. *Ibid.*

28. Paul W. Taylor, *Respect for Nature: A Theory of Environmental Ethics* (Princeton, N.J.: Princeton University Press, 1986).

29. Regan, *The Case for Animal Rights*, p. 153.

30. Michael Allen Fox, *The Case for Animal Experimentation: An Evolutionary and Ethical Perspective* (Berkeley: University of California Press, 1986).

31. Marti Kheel, "The Liberation of Nature: A Circular Affair," *Environmental Ethics*, 7:2 (1985); John Rodman, "Four Forms of Ecological Consciousness Reconsidered," *Ethics and the Environment*, eds. Donald Scherer and Thomas Attig (Englewood Cliffs, N. J.: Prentice-Hall, 1983).

32. Jan Narveson, "On a Case for Animal Rights," *The Monist*, 70: 1 (1987).

33. Jeremy Bentham, "A Utilitarian View," *Animal Rights and Human Obligations*, eds. Tom Regan and Peter Singer (Englewood Cliffs, N. J.: Prentice-Hall, 1976), p. 130.

34. Taylor, *Respect for Nature*, p. 17.

35. Peter Singer, "Animal Liberation or Animal Rights," *The Monist*, 70:1 (1987).

CHAPTER FIVE

1. Bolof Stridbeck, *Ekosofi och Etik* (Göteborg: Bok Skogen, 1994).

2. Angelika Krebs, *Ethics of Nature: Basic Concepts, Basic Arguments of the Present Debate on Animal Ethics and Environmental Ethics* (Ph.D. dissertation, University of Frankfurt, 1993).

3. James Rachels, *Created from Animals: The Moral Implications of Darwinism* (Oxford: Oxford University Press, 1991).

4. E. O. Wilson, ed., *Biodiversity* (Washington: National Academy Press, 1988), and *The Diversity of Life* (London: Penguin Books, 1992).

5. See *Biodiversity, Project Portfolio* (WWF International Publications Unit, 1993), and *The Importance of Biological Diversity*, A Statement by WWF - World Wide Fund For Nature (Gland, Switzerland: WWF International).

6. Arthur O. Lovejoy, *The Great Chain of Being: A Study of the History of an Idea* (London: Harvard University of Press, 1964), p. 313.

7. *Ibid.*, p. 307.

8. *Ibid.*, pp. 294-297.

9. Arthur Schopenhauer, *The Will to Live: Selected Writings of Arthur Schopenhauer*, ed. Richard Taylor (New York: Ungar, 1983), p. 232.

10. *Ibid.*, pp. 199, 206, 217, 219, 221, 232-233.

11. Richard Taylor, "Introduction," in Arthur Schopenhauer, *The Will to Live: Selected Writings of Arthur Schopenhauer*, ed. Richard Taylor (New York: Ungar, 1983), p. xxvii.

12. Schopenhauer, *The Will to Live*, pp. 111, 200.

13. Henri Bergson, *Creative Evolution* (Westport: Greenwood, 1975), pp. 110, 215-217.

14. *Ibid.*, pp. 99, 195.

15. Charles Birch and John B. Cobb, *The Liberation of Life: From the Cell to the Community* (Denton: Environmental Ethics Books, 1990), p. 79.

16. J. E. Lovelock, *Gaia: A New Look at Life on Earth* (Oxford: Oxford University Press, 1982), p. 9.

17. *Ibid.*, pp. 9-11.

18. See Lovejoy, *The Great Chain of Being*, p. 302.

19. Henryk Skolimowski, *Living Philosophy: Eco-Philosophy as a Tree of Life* (London: Arkana, 1992), p. 211.

20. Birch and Cobb, *The Liberation of Life*, pp. 42.

21. *Ibid.*, pp. 91-94.

22. *Ibid.*, pp. 16-17.

23. *Ibid.*

24. *Ibid.*, p. 105.

25. *Ibid.*, pp. 11-12, 42-43.

26. Albert Schweitzer, "The Ethics of Reverence for Life," *Animal Rights and Human Obligations*, eds. Tom Regan and Peter Singer (Englewood Cliffs, N. J.: Prentice-Hall, 1976), p. 133. Albert Schweitzer's *Denken und Tat* is published in Finnish in *Elämän kunnioitus* (Helsinki: WSOY, 1966), pp. 332-350.

27. *Ibid.*, p. 134.

28. Arne Næss, "The Shallow and the Deep, Long-Range Ecology Movement: A Summary," *Inquiry*, 16 (1973), p. 96.

29. Arne Næss, "The Deep Ecological Movement: Some Philosophical Aspects," *Philosophical Inquiry*, 8:1-2 (1986), p. 14.

30. Arne Næss, "Intrinsic Value: Will the Defenders of Nature Please Rise?", *Conservation Biology*, ed. Michael E. Soulé (Sunderland: Sinauer, 1986), p. 507.

31. Arne Næss, *Ecology, Community, and Lifestyle: Outline of an Ecosophy*, trans. and ed. David Rothenberg (Cambridge, England: Cambridge University Press, 1989), p. 29; Bill Devall and George Sessions, *Deep Ecology: Living as if Nature Mattered* (Salt Lake City: Gibbs Smith, 1985), p. 71.

32. Næss, "The Deep Ecological Movement," p. 14; Næss, *Ecology, Community, and Lifestyle*, p. 29; Devall and Sessions, *Deep Ecology*, p. 70.

33. *Ibid.*, p. 67.

34. Næss, "The Deep Ecological Movement," p. 28.

35. Devall and Sessions, *Deep Ecology*, p. 33.

36. Skolimowski, *Living Philosophy*.

37. Henryk Skolimowski, *Eco-Philosophy: Designing New Tactics for Living* (London: Marion Boyars, 1981).

38. Skolimowski, *Living Philosophy*, p. 24.

39. *Ibid*., p. 210.

40. *Ibid*., pp. 25-26.

41. *Ibid*., pp. 103-104.

42. C. I. Lewis, *An Analysis of Knowledge and Valuation* (LaSalle, Ill.: Open Court, 1946).

43. Paul W. Taylor, *Normative Discourse* (Englewood Cliffs, N.J.: Prentice-Hall, 1961), pp. 19, 310, and Taylor, ed., *Problems of Moral Philosophy: An Introduction to Ethics* (Belmont, Cal.: Dickenson, 1967), p. 408; C. A. Baylis, "Grading, Values, and Choice," *Mind*, 67:268 (1958), p. 487; C. I. Lewis, *An Analysis of Knowledge and Valuation* (LaSalle: Open Court, 1946), pp. 382-396.

44. Baylis, "Grading, Values, and Choice," p. 490.

45. Taylor, *Normative Discourse*, p. 23.

46. Paul W. Taylor, "The Ethics of Respect for Nature," *Environmental Ethics*, 3 (1981); Paul W. Taylor, "In Defense of Biocentrism," *Environmental Ethics* (1983); Paul W. Taylor, "Are Humans Superior to Animals and Plants?", *Environmental Ethics*, 6:2 (1984); Paul W. Taylor, *Respect for Nature: A Theory of Environmental Ethics* (Princeton, N.J.: Princeton University Press, 1986).

47. Taylor, *Normative Discourse*; Taylor, ed., *Problems of Moral Philosophy*.

48. Taylor, "Are Humans Superior to Animals and Plants?", *Environmental Ethics*; Taylor, *Respect for Nature*, pp. 71-76.

49. *Ibid*., p. 74.

50. *Ibid*., p. 75.

51. *Ibid*., p. 78.

52. *Ibid*., p. 75.

53. Ernest Partridge, "Values in Nature: Is Anybody There," *Philosophical Inquiry*, 8:1-2 (1986).

54. R. G. Frey, *Interests and Rights: The Case Against Animals* (Oxford: Clarendon Press, 1980).

55. Bryan Norton, "Paul W. Taylor: Respect for Nature," Book Reviews, *Environmental Ethics*, 9:3 (1987), p. 267.

56. Christopher Megone, "R. Attfield: A Theory of Value and Obligation; T. L. S. Sprigge: The Rational Foundations of Ethics," *Journal of Applied Philosophy*, 5:2 (1988), p. 239.

57. Walter E. Howard, "Nature's Role in Animal Welfare," *The Hume Memorial Lecture 29th November 1989* (Universities Federation for Animal Welfare, Potters Bar, 1989), p. 4.

58. Thomas Nagel, "What Is It Like to Be a Bat," *Philosophical Review*, 83 (October 1974).

59. Leena Vilkka, "Miltä tuntuu olla tikankontti," *Näkökulma yhteiskuntatieteelliseen ympäristötutkimukseen*, eds. Terttu Pakarinen, Leena Vilkka, and Eija Moisseinen, Acta Universitatis Tamperensis, B, 37 (Tampere: Tampereen yliopisto, 1991), pp. 48-49.

60. G. H. von Wright, *The Varieties of Goodness* (London: Routledge and Kegan Paul, 1963), p. 51; Taylor, *Respect for Nature*, p. 70.

61. *Ibid.*, p. 69.

62. G. H. Paske, "In Defense of Human 'Chauvinism': A Response to R. Routley and V. Routley," *The Journal of Value Inquiry*, 25:3 (1991), p. 285.

63. *Ibid.*, p. 283.

64. *Ibid.*, p. 284.

65. *Ibid.*, p. 283.

66. Michael Allen Fox, *The Case for Animal Experimentation: An Evolutionary and Ethical Perspective* (Berkeley: University of California Press, 1986), p. 28.

67. von Wright, *The Varieties of Goodness*, p. 50.

68. See R. B. Perry, *General Theory of Value: Its Meaning and Basic Principles Construed in Terms of Interest* (London: Longmans and Green, 1926), p. 50.

69. G. H. von Wright, "Arvot ja tarpeet," *Riittääkö energia, riittääkö järki*, eds. Pentti Malaska, Ismo Kantola, and Pirkko Kasanen (Helsinki: Gaudeamus, 1989), pp. 151-153.

70. Taylor, *Respect for Nature*, pp. 64-65.

71. Tom Regan, *The Case for Animal Rights* (London: Routledge, 1988), p. 87.

CHAPTER SIX

1. Aldo Leopold, *A Sand County Almanac and Sketches Here and There* (Oxford: Oxford University Press, 1968), pp. 224-225.

2. Herbert Marcuse, *One-Dimensional Man* (London: Sphere Books, 1970); Val Plumwood, *Feminism and the Mastery of Nature* (London: Routledge, 1993).

3. Carolyn Merchant, *The Death of Nature: Women, Ecology, and the Scientific Revolution* (London: Harper and Row, 1990).

4. Klaus Michael Meyer-Abich, *Revolution for Nature: From the Environment to the Connatural World*, trans. by Matthew Armstrong (Cambridge, England: The White Horse Press, 1993), p. 28. [*Aufstand für die Natur, Von der Umwelt zur Mitwelt*, 1990], p. 49.

5. *Ibid.*, p. 30, 52.

6. *Ibid.*, p. 30.

7. R. B. Perry, *General Theory of Value: Its Meaning and Basic Principles Construed in Terms of Interest* (London: Longmans and Green, 1926), p. 51.

8. *Ibid.*

9. *Ibid.*, pp. 48-51.

10. Evelyn Pluhar, "Two Conceptions of an Environmental Ethics and Their Implications," *Ethics and Animals*, 4 (1983), p. 123.

11. Andrew Brennan, "Moral Pluralism and the Environment," *Environmental Values*, 1:1 (1992), p. 25.

12. S. F. Sapontzis, *Moral, Reason, and Animals* (Philadelphia: Temple University Press, 1987), p. 271.

13. J. Baird Callicott, *In Defense of the Land Ethic: Essays in Environmental Philosophy* (Albany: State University of New York Press, 1989), p. 72.

14. Lawrence E. Johnson, *A Morally Deep World: An Essay on Moral Significance and Environmental Ethics* (Cambridge, Mass.: Cambridge University Press, 1991), p. 211, 267.

15. Meyer-Abich, *Revolution for Nature*, p. 29.

16. See *ibid.*, pp. 135, 74.

17. *Ibid.*

18. Leopold, *A Sand County Almanac and Sketches Here and There*, p. viii.

19. *Ibid.*, p. 216.

20. *Ibid.*, p. 204.

21. *Ibid.*

22. *Ibid.*, p. ix.

23. Susan L. Flader, *Thinking Like a Mountain: Aldo Leopold and the Evolution of an Ecological Attitude toward Deer, Wolves, and Forests* (Columbia, Mo.: University of Missouri Press, 1974), p. 1.

24. Leopold, *A Sand County Almanac and Sketches Here and There*, pp. 202-203.

25. *Ibid.*, p. 203.

26. Andrew Brennan, *Thinking about Nature: An Investigation of Nature, Value and Ecology* (London: Routledge, 1988), p. 217.

27. *Ibid.*

28. David E. Cooper, "The Idea of Environment," *The Environment in Question*, eds. David E. Cooper and Joy A. Palmer (London: Routledge, 1992), p. 167.

29. Eugene C. Hargrove, *Foundations of Environmental Ethics* (Englewood Cliffs, N. J.: Prentice-Hall, 1989).

30. Yrjö Sepänmaa, *The Beauty of Environment: A General Model for Environmental Aesthetics* (Denton, Texas: Environmental Ethics Books, 1993).

31. Holmes Rolston, III, *Conserving Natural Value* (New York: Columbia University Press, 1994), p. 136.

32. Jared M. Diamond, "Evolution of Bowerbirds' Bowers: Animal Origins of the Aesthetic Sense," *Nature*, 297 (13 May 1982), p. 99.

33. Edgar S. Brightman, *Nature and Values* (New York: Abingdon-Cokesbury Press, 1945), p. 71.

34. Robert S. Hartman, "The Nature of Valuation," in Edwards and Davis, eds., *Forms of Value and Valuation*, p. 31.

35. G. E. Moore, *Principia Ethica* (Cambridge, England: Cambridge University Press, 1988), pp. xi, 28.

36. Ralph Waldo Emerson and Henry David Thoreau, *Nature/Walking* (Boston: Beacon Press, 1991).

37. Hargrove, *Foundations of Environmental Ethics.*

38. *Ibid.*, p. 198.

39. G. E. Moore, *Philosophical Studies* (London: Routledge and Kegan Paul, 1951), p. 255.

40. *Ibid.*, p. 260.

41. *Ibid.*, p. 261.

42. See Jaegwon Kim, "Concepts of Supervenience," *Philosophy and Phenomenological Research*, 45:2 (1984).

43. Sepänmaa, *The Beauty of Environment*. See also Jack L. Nasar, *Environmental Aesthetics: Theory, Research, and Applications* (Cambridge, England: Cambridge University Press, 1992); Martin Seel, *Eine Ästhetik der Natur* (Frankfurt am Main: Suhrkamp, 1991); Cheryl Foster, *Aesthetics and the Natural Environment* (Ph.D. dissertation, University of Edinburgh, 1993).

44. Henryk Skolimowski, *Living Philosophy: Eco-Philosophy as a Tree of Life* (London: Arkana, 1992), pp. 25-26.

45. Cf. *ibid.*, pp. 103-104.

CHAPTER SEVEN

1. Yrjö Haila and Richard Levins, *Humanity and Nature: Ecology, Science, and Society* (London: Pluto Press, 1992).

2. Timo Airaksinen, "Vielä kerran: Arvojen ja tosiasioiden suhde," *Epäilyttäviä esseitä: S. Albert Kivisen 60-vuotispäivän kunniaksi*, eds. Tuomo Aho and Gabriel Sandu (Helsinki: Helsingin yliopiston filosofian laitoksen julkaisuja, 1993).

3. Risieri Frondizi, *What Is Value? An Indroduction to Axiology* (LaSalle, Ill.: Open Court, 1971), p. ix.

4. *Ibid.*, p. 19.

5. *Ibid.*, p. 18.

6. *Ibid.*, p. 123.

7. *Ibid.*, p. 123.

8. See Robyn Eckersley, *Environmentalism and Political Theory: Toward an Ecocentric Approach* (London: UCL Press, 1992), p. 55.

9. *Ibid.*

10. Frondizi, *What Is Value?*, p. 20.

11. Andrew Dobson, *Green Political Thought* (London: Unwin Hyman, 1990), p. 53.

12. Laura Westra, *An Environmental Proposal for Ethics: The Principle of Integrity* (London: Rowman and Littlefield, 1994), p. 124.

13. Lawrence E. Johnson, *A Morally Deep World: An Essay on Moral Significance and Environmental Ethics* (Cambridge, Mass.: Cambridge University Press, 1991).

14. J. V. Krutilla, "Conservation Reconsidered," *American Economic Review*, 57 (1967).

15. Alan Randall, "A Total Value Framework for Benefit Estimation," *Valuing Wildlife Resources in Alaska*, eds. George L. Peterson, Cindy Sorg Swanson, Daniel W. McCollum, and Michael H. Thomas (Oxford: Westview, 1992).

CHAPTER EIGHT

1. Yrjö Haila and Richard Levins, *Humanity and Nature: Ecology, Science, and Society* (London: Pluto Press, 1992).

2. Rem B. Edwards, "Universals, Individuals, and Intrinsic Goods," in *Forms of Value and Valuation: Theory and Applications*, Rem B. Edwards and John W. Davis (London: University Press of America, 1991).

3. Paul W. Taylor, *Respect for Nature: A Theory of Environmental Ethics* (Princeton, N.J.: Princeton University Press, 1986).

4. Andrew Brennan, *Thinking about Nature: An Investigation of Nature, Value and Ecology* (London: Routledge, 1988).

5. A. T. Nuyen, "An Anthropocentric Ethics Towards Animals and Nature," *The Journal of Value Inquiry*, 15 (1981).

6. Bryan Norton, *Why Preserve Natural Variety?* (Princeton, N.J.: Princeton University Press, 1987).

7. Brennan, *Thinking about Nature.*

8. Eugene C. Hargrove, "Weak Anthropocentric Intrinsic Value," *The Monist,* 75:2 (1992).

9. John O'Neill, *Ecology, Policy, and Politics* (London: Routledge, 1993).

10. David Ehrenfeld, *The Arrogance of Humanism* (Oxford: Oxford University Press, 1981).

11. Robert M. May, "The Modern Biologist's View of Nature," in John Torrance, ed., *The Concept of Nature*, The Herbert Spencer Lectures (Oxford: Clarendon Press, 1992).

12. G. H. von Wright, *Humanismi elämänasenteena* (Helsinki: Otava, 1983).

13. Henryk Skolimowski, *Eco-Philosophy: Designing New Tactics for Living* (London: Marion Boyars, 1981).

14. Brennan, *Thinking about Nature.*

15. Christine M. Korsgaard, "Two Distinctions in Goodness," *The Philosophical Review*, 92:2 (April 1983).

16. Henryk Skolimowski, *Living Philosophy: Eco-Philosophy as a Tree of Life* (London: Arkana, 1992).

17. Ehrenfeld, *The Arrogance of Humanism*, p. 21.

18. Korsgaard, "Two Distinctions in Goodness."

CHAPTER NINE

1. Paul W. Taylor, *Respect for Nature: A Theory of Environmental Ethics* (Princeton, N. J.: Princeton University Press, 1986) and Taylor, "Inherent Value and Moral Rights," *The Monist*, 70:1 (1987).

2. Roderick Frazier Nash, *The Rights of Nature: A History of Environmental Ethics* (London: The University of Wisconsin Press, 1989).

3. Bernard Gendron, "The Viability of Environmental Ethics," *Philosophy and Technology*, eds. Paul T. Durbin and Friedrich Rapp (Amsterdam: Reidel, 1983), p. 192.

4. *Ibid.*

5. T. L. S. Sprigge, "Non-Human Rights: An Idealist Perspective," *Inquiry*, 27 (1984).

6. Louis G. Lombardi, "Inherent Worth, Respect, and Rights," *Environmental Ethics*, 5 (Fall 1983).

7. John Passmore, *Man's Responsibility for Nature: Ecological Problems and Western Traditions* (London: Duckworth, 1980), p. 116.

8. R. G. Frey, *Interests and Rights: The Case Against Animals* (Oxford: Clarendon Press, 1980).

9. Cf. Eugene C. Hargrove, ed, *Religion and Environmental Crisis* (London: The University of Georgia Press, 1986).

10. See Tom Regan, *The Case for Animal Rights* (London: Routledge, 1988), and Mary Anne Warren, "The Rights of the Nonhuman World," in Robert Elliot and Arran Gare, eds., *Environmental Philosophy: A Collection of Readings* (Milton Keynes: The Open University Press, 1983).

11. Charles R. Magel, *Keyguide to Information Sources in Animal Rights* (London: Mansell, 1989).

12. "Animal Rights," *Inquiry*, 22:1-2 (1979); "Animal Rights," *The Monist*, 70:1 (1987).

13. Peter Singer, *Animal Liberation: Towards an End to Man's Inhumanity to Animals* (London: Granada, 1977); Regan, *The Case for Animal Rights*; Michael W. Fox, *Inhumane Society: The American Way of Exploiting Animals* (New York: St. Martin's Press, 1990); Richard Ryder, *Victims of Science: The Use of Animals in Research* (London: National Anti-Vivisection Society, 1983); Mary Midgley, *Animals and Why They Matter* (Athens, Ga.: The University of Georgia Press, 1984); S. F. Sapontzis, *Moral, Reason, and Animals* (Philadelphia: Temple University Press, 1987); Bernard Rollin, *The Unheeded Cry: Animal Consciousness, Animal Pain, and Science* (Oxford University Press, 1989).

14. R. G. Frey, *Rights, Killing, and Suffering: Moral Vegetarianism and Applied Ethics* (Oxford: Basil Blackwell, 1983).

15. Bernard Rollin, *Animal Rights and Human Morality* (New York: Prometheus Books, 1981).

16. Leena Vilkka, "A New Dimension of Animal Rights," *Satya* (May 1995).

17. Stephen Clark, "Animals, Ecosystems, and the Liberal Ethic," *The Monist*, 70:1 (1987).

18. James Rachels, *Created from Animals: The Moral Implications of Darwinism* (Oxford: Oxford University Press, 1991).

19. *A Declaration on Great Apes, The Great Ape Project*, PO Box 1023, Collingwood, Melbourne, Victoria, Australia 3066.

20. Nicholas Rescher, *Unpopular Essays on Technological Progress* (Pittsburgh: University of Pittsburgh Press, 1980).

21. Joel Feinberg, "The Rights of Animals and Unborn Generations," in William T. Blackstone, ed., *Philosophy and Environmental Crisis* (Athens, Ga.: University of Georgia Press, 1974), p. 51.

22. *Ibid.*

23. Bryan Norton, "Environmental Ethics and Nonhuman Rights," *Environmental Ethics*, 4 (Spring 1982), pp. 24-25.

24. Scott Lehmann, "Do Wildernesses Have Rights," *Environmental Ethics*, 3:2 (1981).

25. Gendron, "The Viability of Environmental Ethics," p. 194.

26. William T. Blackstone, ed., *Philosophy and Environmental Crisis* (Athens, Ga.: University of Georgia Press, 1974).

27. Eric Ashby, "Towards an Environmental Ethic," *Nature*, 262:8 (July 1976).

28. Cristopher Stone, *Should Trees Have Standing: Toward Legal Rights for Natural Objects* (Los Altos, Cal.: William Kaufmann, 1974).

29. Blackstone, ed., *Philosophy and Environmental Crisis*.

30. Stone, *Should Trees Have Standing*.

31. Arne Næss, *Ecology, Community, and Lifestyle: Outline of an Ecosophy*, trans. and ed. by David Rothenberg (Cambridge, England: Cambridge University Press, 1989).

32. William T. Blackstone, "Ecology and Rights," *Environmental Ethics*, ed. Kristin S. Shrader-Frechette (Boxwood, Pacific Grove, 1981).

33. Hargrove, ed., *Religion and Environmental Crisis*.

BIBLIOGRAPHY

Adorno, Theodor and Max Horkheimer. *Dialectic of Enlightenment* (London: Verso Editions, 1979) [*Dialektik der Aufklärung*, 1944].

Aho, Tuomo and Gabriel Sandu, eds. *Epäilyttäviä esseitä: S. Albert Kivisen 60-vuotispäivän kunniaksi* (Helsinki: Helsingin yliopiston filosofian laitos, 1993).

Airaksinen, Timo. "Vielä kerran: Arvojen ja tosiasioiden suhde," *Epäilyttäviä esseitä: S. Albert Kivisen 60-vuotispäivän kunniaksi*, eds. Tuomo Aho and Gabriel Sandu (Helsinki: Helsingin yliopiston filosofian laitos, 1993).

"Animal Rights," *Inquiry*, 22:1-2 (1979).

"Animal Rights," *The Monist*, 70:1 (1987).

Aquinas, Thomas. "Differences between Rational and Other Creatures," *Animal Rights and Human Obligations*, eds. Tom Regan and Peter Singer (Englewood Cliffs, N. J.: Prentice-Hall, 1976).

Ashby, Eric. "Towards an Environmental Ethic," *Nature*, 262:8 (July 1976).

Bahm, Archie J. *Axiology: The Science of Values* (Amsterdam and Atlanta: Rodopi, Value Inquiry Book Series, 1993).

Baylis, C. A. "Grading, Values, and Choice," *Mind*, 67:268 (1958).

Bentham, Jeremy. "A Utilitarian View," *Animal Rights and Human Obligations*, eds. Tom Regan and Peter Singer (Englewood Cliffs, N. J.: Prentice-Hall, 1976).

Bergson, Henri. *Creative Evolution* (Westport: Greenwood, 1975) [1944].

Biodiversity, Project Portfolio (Gland, Switzerland: WWF International Publications Unit, 1993).

Birch, Charles and John B. Cobb. *The Liberation of Life: From the Cell to the Community* (Denton: Environmental Ethics Books, 1990) [1981].

Blackstone, William T., ed. *Philosophy and Environmental Crisis* (Athens, Ga.: University of Georgia Press, 1974).

———. "Ecology and Rights," *Environmental Ethics*, ed. Kristin S. Shrader-Frechette (Boxwood, Pacific Grove, 1981).

Brennan, Andrew. *Thinking about Nature: An Investigation of Nature, Value, and Ecology* (London: Routledge, 1988).

———. "Moral Pluralism and the Environment," *Environmental Values*, 1:1 (1992).

Brightman, Edgar S. *Nature and Values* (New York: Abingdon-Cokesbury Press, 1945).

Broom, Donald and K. G. Johnson. *Stress and Animal Welfare* (London: Chapman & Hall, 1993).

Callicott, J. Baird. "On the Intrinsic Value of Nonhuman Species," *The Preservation of Species: The Value of Biological Diversity*, ed. Bryan Norton (Princeton, N.Y.: Princeton University Press, 1986).

———. *In Defense of the Land Ethic: Essays in Environmental Philosophy* (Albany: State University of New York Press, 1989).

Caplan, Arthur L., H. Tristram Engelhardt, Jr., and James J. McCartney, eds. *Concepts of Health and Disease: Interdisciplinary Perspectives* (London: Addison-Wesley, 1981).

Carruthers, Peter. *The Animals Issue: Moral Theory in Practice* (Cambridge, England: Cambridge University Press, 1992).

Carter, Robert E. "The Norm of Intrinsic Valuation," *Forms of Value and Valuation: Theory and Applications*, Rem B. Edwards and John W. Davis (London: University Press of America, 1991).

Clark, Stephen. "Animals, Ecosystems, and the Liberal Ethic," *The Monist*, 70:1 (1987).

Close, Bryony, Francine Dolins, and Georgia Mason, eds. *Euroniche Conference Proceedings: Animal Use in Education* (London: Humane Education Centre, 1989).

Cooper, David E. "The Idea of Environment," *The Environment in Question*, eds. David E. Cooper and Joy A. Palmer (London: Routledge, 1992).

Cooper, David E. and Joy A. Palmer, eds. *The Environment in Question* (London: Routledge, 1992).

Dawkins, Marian Stamp. "The Scientific Basis for Assessing Suffering in Animals," *In Defence of Animals*, Peter Singer (Oxford: Basil Blackwell, 1986) [1985].

———. "From an Animal's Point of View: Motivation, Fitness, and Animal Welfare," *Behavioral and Brain Sciences*, 13 (1990).

A Declaration on Great Apes, The Great Ape Project, PO Box 1023, Collingwood, Melbourne, Victoria, Australia 3066.

Devall, Bill and George Sessions. *Deep Ecology: Living as if Nature Mattered* (Salt Lake City: Gibbs Smith, 1985).

Diamond, Jared M. "Evolution of Bowerbirds' Bowers: Animal Origins of the Aesthetic Sense," *Nature*, 297 (13 May 1982).

Dobson, Andrew. *Green Political Thought* (London: Unwin Hyman, 1990).

Dubois, R. F. *Loup, Mon Ami. Wolf, My Friend* (Outrelouxhe, Belgium: Wolves and Wildlife Society, 1994).

Durbin, Paul T. and Friedrich Rapp, eds. *Philosophy and Technology* (Amsterdam: Reidel, 1983).

Eckersley, Robyn. *Environmentalism and Political Theory: Toward an Ecocentric Approach* (London: UCL Press, 1992).

Edwards, Rem B. "Universals, Individuals, and Intrinsic Goods," *Forms of Value and Valuation: Theory and Applications*, Rem B. Edwards and John W. Davis (London: University Press of America, 1991).

Edwards, Rem B. and John W. Davis, eds. *Forms of Value and Valuation: Theory and Applications* (London: University Press of America, 1991).

Ehrenfeld, David. *The Arrogance of Humanism* (Oxford: Oxford University Press, 1981).

Elliot, Robert. "Intrinsic Value, Environmental Obligation, and Naturalness," *The Monist*, 75: 2 (1992).

Elliot, Robert and Arran Gare, eds. *Environmental Philosophy: A Collection of Readings* (The Open University Press, Milton Keynes, 1983).

Ellul, Jacques. *The Technological Society*, trans. by John Wilkinson (New York: Vintage Books, 1964).

Emerson, Ralph Waldo and Henry David Thoreau, *Nature/Walking* (Boston: Beacon Press, 1991).

Feinberg, Joel. "The Rights of Animals and Unborn Generations," *Philosophy and Environmental Crisis*, ed. William T. Blackstone (Athens, Ga.: University of Georgia Press, 1974).

Flader, Susan L. *Thinking Like a Mountain: Aldo Leopold and the Evolution of an Ecological Attitude toward Deer, Wolves, and Forests* (Columbia, Mo.: University of Missouri Press, 1974).

Foster, Cheryl. *Aesthetics and the Natural Environment* (Ph.D. dissertation, University of Edinburgh, 1993).

Fox, Michael Allen. *The Case for Animal Experimentation: An Evolutionary and Ethical Perspective* (Berkeley: University of California Press, 1986).

Fox, Michael W. *Inhumane Society: The American Way of Exploiting Animals* (New York: St. Martin's Press, 1990).

Frankena, William K. *Ethics* (Englewood Cliffs, N. J.: Prentice-Hall, 1973) [1963].

Frey, R. G. *Interests and Rights: The Case Against Animals* (Oxford: Clarendon Press, 1980).

———. *Rights, Killing, and Suffering: Moral Vegetarianism and Applied Ethics* (Oxford: Basil Blackwell, 1983).

Frondizi, Risieri. *What Is Value? An Introduction to Axiology* (LaSalle, Ill.: Open Court, 1971).

Gendron, Bernard. "The Viability of Environmental Ethics," *Philosophy and Technology*, eds. Paul T. Durbin and Friedrich Rapp (Amsterdam: Reidel, 1983).

Giere, Ronald N. *Understanding Scientific Reasoning* (Orlando, Fla.: Holt, Rinehart, and Winston, Dryden Press, 1991).

Goodpaster, Kenneth E. "On Being Morally Considerable," *Ethics and the Environment,* eds. Donald Scherer and Thomas Attig (Englewood Cliffs, N.J.: Prentice-Hall, 1983).

Gracia, Jorge J. E. "Evil and the Transcendentality of Goodness: Suárez's Solution to the Problem of Positive Evils," *Being and Goodness: The Concept of the Good in Metaphysics and Philosophical Theology*, ed. Scott MacDonald (Ithaca, N.Y.: Cornell University Press, 1991).

Haila, Yrjö and Richard Levins. *Humanity and Nature: Ecology, Science, and Society* (London: Pluto Press, 1992).

Hargrove, Eugene C., ed. *Religion and Environmental Crisis* (London: The University of Georgia Press, 1986).

———. *Foundations of Environmental Ethics* (Englewood Cliffs, N.J.: Prentice-Hall, 1989).

———. "Weak Anthropocentric Intrinsic Value," *The Monist*, 75:2 (1992).

Harman, Gilbert H. "Toward a Theory of Intrinsic Value," *The Journal of Philosophy*, 64 (1967).

Hartman, Robert S. "The Nature of Valuation," *Forms of Value and Valuation: Theory and Applications*, Rem B. Edwards and John W. Davis (London: University Press of America, 1991).

Hooker, C. A. "Responsibility, Ethics, and Nature," *The Environment in Question*, eds. David E. Cooper and Joy A. Palmer (London: Routledge, 1992).

Howard, Walter E. "Nature's Role in Animal Welfare," *The Hume Memorial Lecture 29th November 1989* (London: Universities Federation for Animal Welfare, 1989).

The Importance of Biological Diversity, A Statement by WWF - World Wide Fund For Nature (Gland, Switzerland: WWF International, 1986).

Johnson, Lawrence E. *A Morally Deep World: An Essay on Moral Significance and Environmental Ethics* (Cambridge, Mass.: Cambridge University Press, 1991).

Kant, Immanuel. "Duties to Animals," *Animal Rights and Human Obligations*, eds. Tom Regan and Peter Singer (Englewood Cliffs, N. J.: Prentice-Hall, 1976).

———. *Ethical Philosophy*, The complete texts of *Grounding for the Metaphysics of Morals* [1785] and *Metaphysical Principles of Virtue* [1797], trans. by James W. Ellington (Indianapolis, Ind.: Hackett, 1983).

Katz, Eric. "Searching for Intrinsic Value," *Environmental Ethics*, 9:3 (1987).

Kennedy, J. S. *The New Anthropomorphism* (Cambridge, England: Cambridge University Press, 1992).

Kheel, Marti. "The Liberation of Nature: A Circular Affair," *Environmental Ethics*, 7:2 (1985)

Kim, Jaegwon. "Concepts of Supervenience," *Philosophy and Phenomenological Research*, 45:2 (1984).

Kohák, Erazim. "Why Is There Something Good, Not Simply Something?", *The Journal of Speculative Philosophy*, 5:1 (1991).

Korsgaard, Christine M. "Two Distinctions in Goodness," *The Philosophical Review*, 92:2 (April 1983).

Krebs, Angelika. *Ethics of Nature: Basic Concepts, Basic Arguments of the Present Debate on Animal Ethics and Environmental Ethics* (Ph.D. dissertation, University of Frankfurt, 1993).

Krutilla, J. V. "Conservation Reconsidered," *American Economic Review*, 57 (1967).

Lachs, John. "Is Everything Intrinsically Good?", *The Journal of Speculative Philosophy*, 5:1 (1991).

Leahy, Michael. *Against Liberation: Putting Animals in Perspective* (London: Routledge, 1991).

Lehmann, Scott. "Do Wildernesses Have Rights," *Environmental Ethics*, 3:2 (1981).

Leopold, Aldo. *A Sand County Almanac and Sketches Here and There* (Oxford: Oxford University Press, 1968).

Lewis, C. I. *An Analysis of Knowledge and Valuation* (LaSalle, Ill.: Open Court, 1946).

Lombardi, Louis G. "Inherent Worth, Respect, and Rights," *Environmental Ethics*, 5 (Fall 1983)

Lovejoy, Arthur O. *The Great Chain of Being: A Study of the History of an Idea* (London: Harvard University Press, 1964) [1936].

Lovelock, J. E. *Gaia: A New Look at Life on Earth* (Oxford: Oxford University Press, 1982) [1979].

MacDonald, Scott, ed. *Being and Goodness: The Concept of the Good in Metaphysics and Philosophical Theology* (Ithaca, N.Y.: Cornell University Press, 1991).

Magel, Charles R. *Keyguide to Information Sources in Animal Rights* (London: Mansell, 1989).

Malaska, Pentti, Ismo Kantola, and Pirkko Kasanen, eds. *Riittääkö energia, riittääkö järki* (Helsinki: Gaudeamus, 1989).

Marcuse, Herbert. *One-Dimensional Man* (London: Sphere Books, 1970).

May, Robert M. "The Modern Biologist's View of Nature," *The Concept of Nature*, ed. John Torrance, The Herbert Spencer Lectures (Oxford: Clarendon Press, 1992).

Megone, Christopher. "R. Attfield: A Theory of Value and Obligation; T. L. S. Sprigge: The Rational Foundations of Ethics," book reviews in *Journal of Applied Philosophy*, 5:2 (1988).

Merchant, Carolyn. *The Death of Nature: Women, Ecology, and the Scientific Revolution* (London: Harper and Row, 1990) [1980].

Meyer-Abich, Klaus Michael. *Revolution for Nature: From the Environment to the Connatural World*, trans. by Matthew Armstrong (Cambridge, England: The White Horse Press, 1993) [*Aufstand für die Natur, Von der Umwelt zur Mitwelt*, 1990].

Midgley, Mary. *Animals and Why They Matter* (Athens, Ga.: The University of Georgia Press, 1984).

The Monist, General Topic: The Intrinsic Value of Nature, An International Journal of General Philosophical Inquiry, 75:2 (1992).

Moore, G. E. *Philosophical Studies* (London: Routledge and Kegan Paul, 1951) [1922].

———. *Principia Ethica* (Cambridge, England: Cambridge University Press, 1988) [1903].

Mumford, Lewis. *The Myth of the Machine: Technics and Human Development* (New York: Harcourt Brace Jovanovich, 1966).

Næss, Arne. "The Shallow and the Deep, Long-Range Ecology Movement: A Summary," *Inquiry*, 16 (1973).

———. "The Deep Ecological Movement: Some Philosophical Aspects," *Philosophical Inquiry*, 8:1-2 (1986).

———. "Intrinsic Value: Will the Defenders of Nature Please Rise?", *Conservation Biology*, ed. Michael E. Soulé (Sunderland: Sinauer, 1986).

———. *Ecology, Community, and Lifestyle: Outline of an Ecosophy*, trans. and ed. David Rothenberg (Cambridge, England: Cambridge University Press, 1989).

Nagel, Thomas. "What Is It Like to Be a Bat," *Philosophical Review*, 83 (October 1974).

Narveson, Jan. "On a Case for Animal Rights," *The Monist*, 70:1 (1987).

Nasar, Jack L. *Environmental Aesthetics: Theory, Research, and Applications* (Cambridge, England: Cambridge University Press, 1992) [1988].

Nash, Roderick Frazier. *The Rights of Nature: A History of Environmental Ethics* (London: The University of Wisconsin Press, 1989).

Norton, Bryan. "Environmental Ethics and Nonhuman Rights," *Environmental Ethics*, 4 (Spring 1982).

———. *Why Preserve Natural Variety?* (Princeton: Princeton University Press, 1987).

———. "Paul W. Taylor: Respect for Nature," Book Reviews, *Environmental Ethics*, 9:3 (1987).

———, ed. *The Preservation of Species: The Value of Biological Diversity* (Princeton: Princeton University Press, 1986).

Nuyen, A. T. "An Anthropocentric Ethics Towards Animals and Nature," *The Journal of Value Inquiry*, 15 (1981).

O'Neill, John. *Ecology, Policy, and Politics* (London: Routledge, 1993).

Pakarinen, Terttu, Leena Vilkka, and Eija Moisseinen, eds. *Näkökulma yhteiskuntatieteelliseen ympäristötutkimukseen*, Acta Universitatis Tamperensis, ser. B, 37 (Tampere: Tampereen yliopisto, 1991).

Paloheimo, Eero. *Maan tie* (Helsinki: WSOY, 1989).

Partridge, Ernest. "Values in Nature: Is Anybody There," *Philosophical Inquiry*, 8:1-2 (1986).

Paske, G. H. "In Defense of Human 'Chauvinism': A Response to R. Routley and V. Routley," *The Journal of Value Inquiry*, 25:3 (1991).

Passmore, John. *Man's Responsibility for Nature: Ecological Problems and Western Traditions* (London: Duckworth, 1980) [1974].

Paterson, David and Mary Palmer, eds. *The Status of Animals: Ethics, Education, and Welfare* (Wallingford: Humane Education Foundation, 1989).

Perry, R. B. *General Theory of Value: Its Meaning and Basic Principles Construed in Terms of Interest* (London: Longmans and Green, 1926).

Peterson, George L., Cindy Sorg Swanson, Daniel W. McCollum, and Michael H. Thomas, eds. *Valuing Wildlife Resources in Alaska* (Oxford: Westview, 1992).

Pluhar, Evelyn. "Two Conceptions of an Environmental Ethics and Their Implications," *Ethics and Animals*, 4 (1983).

Plumwood, Val. *Feminism and the Mastery of Nature* (London: Routledge, 1993).

Pylkkänen, Paavo and Pauli Pylkkö, eds. *New Directions in Cognitive Science* (Helsinki: Finnish Artificial Intelligence Society, 1995).

Rachels, James. *Created from Animals: The Moral Implications of Darwinism* (Oxford: Oxford University Press, 1991) [1990].

Randall, Alan. "A Total Value Framework for Benefit Estimation," *Valuing Wildlife Resources in Alaska*, eds. George L. Peterson, Cindy Sorg Swanson, Daniel W. McCollum, and Michael H. Thomas (Oxford: Westview, 1992).

Regan, Tom. "The Nature and Possibility of an Environmental Ethic," *Environmental Ethics*, 3:1 (1981).

———. *The Case for Animal Rights* (London: Routledge, 1988) [1984].

Regan, Tom and Peter Singer, eds. *Animal Rights and Human Obligations* (Englewood Cliffs, N.J.: Prentice-Hall, 1976).

Rescher, Nicholas. *Unpopular Essays on Technological Progress* (Pittsburgh: University of Pittsburgh Press, 1980).

Rodman, John. "Four Forms of Ecological Consciousness Reconsidered," *Ethics and the Environment*, eds. Donald Scherer and Thomas Attig (Englewood Cliffs, N.J.: Prentice-Hall, 1983).

Rollin, Bernard. *Animal Rights and Human Morality* (New York: Prometheus Books, 1981).

———. "The Teaching of Responsibility," *The Hume Memorial Lecture: 10th November 1983 at King's College* (London: University of London, UFAW, 1983).

———. *The Unheeded Cry: Animal Consciousness, Animal Pain, and Science* (Oxford: Oxford University Press, 1989).

Rolston, Holmes, III. *Conserving Natural Value* (New York: Columbia University Press, 1994).

Ryder, Richard. *Victims of Science: The Use of Animals in Research* (London: National Anti-Vivisection Society, 1983) [1975].

Sapontzis, S. F. *Morals, Reason, and Animals* (Philadelphia: Temple University Press, 1987).

———. "Animal Rights and Biomedical Research," *The Journal of Value Inquiry*, 26:1 (1992).

Scherer, Donald and Thomas Attig, eds. *Ethics and the Environment* (Englewood Cliffs, N.J.: Prentice-Hall, 1983).

Schopenhauer, Arthur. *The Will to Live: Selected Writings of Arthur Schopenhauer*, ed. Richard Taylor (New York: Ungar, 1983) [1967].

Schumacher, E. F. *Small Is Beautiful: Economics as If People Mattered* (New York: Harper and Row, 1973).

Schweitzer, Albert. *Elämän kunnioitus*, trans. by Juho Tervonen (Helsinki: WSOY, 1966) [*Denken und Tat*].

———. "The Ethics of Reverence for Life," *Animal Rights and Human Obligations*, eds. Tom Regan and Peter Singer (Englewood Cliffs, N. J.: Prentice-Hall, 1976).

Scoville, Judith. "Value Theory and Ecology in Environmental Ethics," *Environmental Ethics*, 17:2 (1995).

Seel, Martin. *Eine Ästhetik der Natur*, Suhrkamp (Frankfurt am Mein: 1991).

Sepänmaa, Yrjö. *The Beauty of Environment: A General Model for Environmental Aesthetics* (Denton, Texas: Environmental Ethics Books, 1993).

Shrader-Frechette, Kristin S., ed. *Environmental Ethics* (Pacific Grove, Cal.: Boxwood, 1981).

Singer, Peter. *Animal Liberation: Towards an End to Man's Inhumanity to Animals* (London: Granada, 1977) [1975].

———. *Practical Ethics* (Cambridge, England: Cambridge University Press, 1986) [1979].

———. "Animal Liberation or Animal Rights," *The Monist*, 70:1 (1987).

———, ed. *In Defence of Animals* (Oxford: Basil Blackwell, 1986) [1985].

Skolimowski, Henryk. *Eco-Philosophy: Designing New Tactics for Living* (London: Marion Boyars, 1981).

———. *Living Philosophy: Eco-Philosophy as a Tree of Life* (London: Arkana, 1992).

Society and Animals, Social Scientific Studies of the Human Experience of Other Animals, ed. Kenneth Shapiro, assoc. ed. Arnold Arluke (Cambridge, England: The White Horse Press).

Sprigge, T. L. S. "Non-Human Rights: An Idealist Perspective," *Inquiry*, 27 (1984).

———. "The Rational Foundations of Ethics," *Journal of Applied Philosophy*, 5:2 (1988).

Stevenson, Charles L. *Ethics and Language* (New Haven: Yale University Press, 1947) [1945].

Stewart, M. F. "Teaching of Animal Welfare to Veterinary Students," *The Status of Animals: Ethics, Education, and Welfare*, eds. David Paterson and Mary Palmer (Wallingford: Humane Education Foundation, 1989).

Stone, Cristopher. *Should Trees Have Standing: Toward Legal Rights for Natural Objects* (Los Altos, Cal.: William Kaufmann, 1974).

Stridbeck, Bolof. *Ekosofi och Etik* (Göteborg: Bok Skogen, 1994).

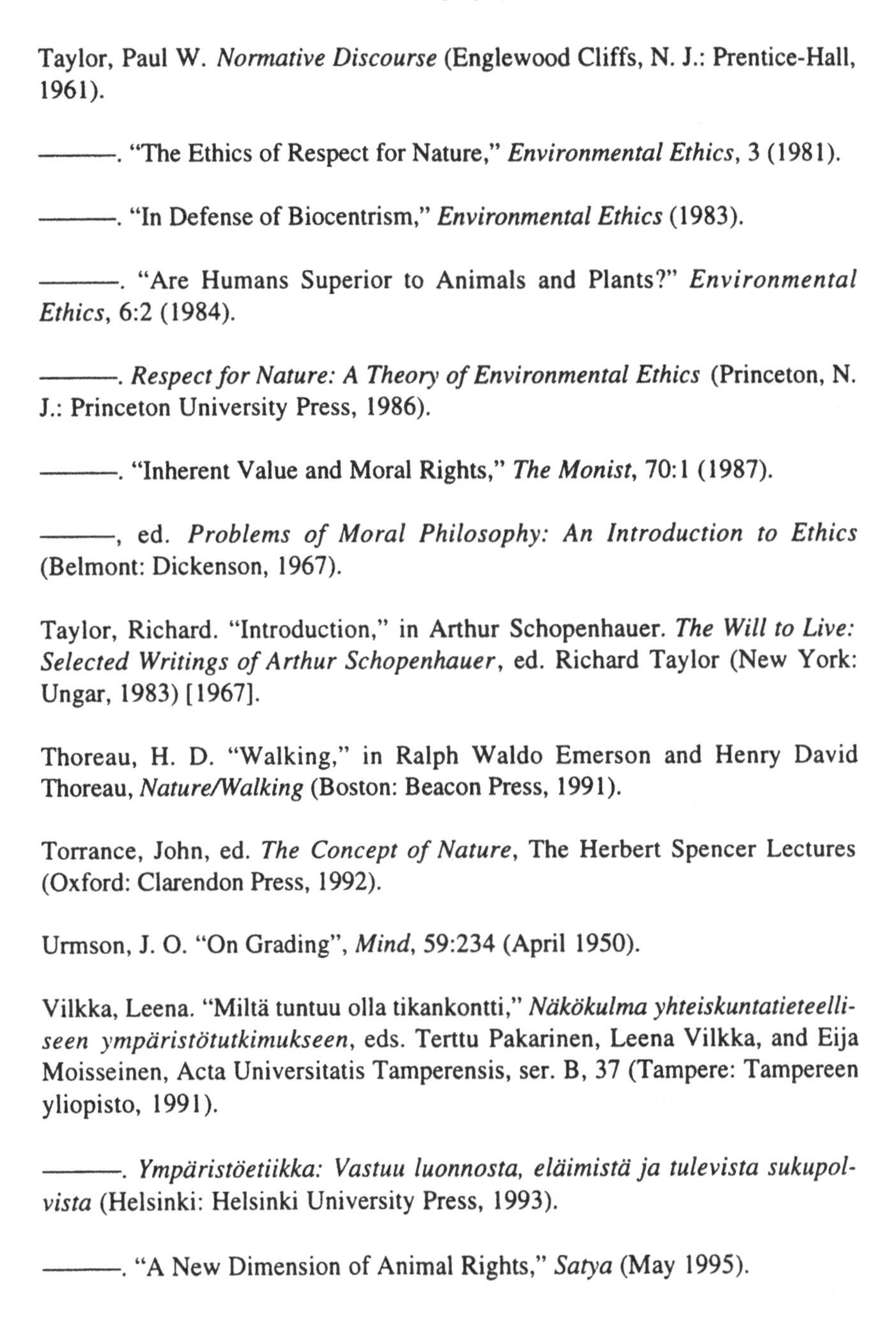

Taylor, Paul W. *Normative Discourse* (Englewood Cliffs, N. J.: Prentice-Hall, 1961).

———. "The Ethics of Respect for Nature," *Environmental Ethics*, 3 (1981).

———. "In Defense of Biocentrism," *Environmental Ethics* (1983).

———. "Are Humans Superior to Animals and Plants?" *Environmental Ethics*, 6:2 (1984).

———. *Respect for Nature: A Theory of Environmental Ethics* (Princeton, N. J.: Princeton University Press, 1986).

———. "Inherent Value and Moral Rights," *The Monist*, 70:1 (1987).

———, ed. *Problems of Moral Philosophy: An Introduction to Ethics* (Belmont: Dickenson, 1967).

Taylor, Richard. "Introduction," in Arthur Schopenhauer. *The Will to Live: Selected Writings of Arthur Schopenhauer*, ed. Richard Taylor (New York: Ungar, 1983) [1967].

Thoreau, H. D. "Walking," in Ralph Waldo Emerson and Henry David Thoreau, *Nature/Walking* (Boston: Beacon Press, 1991).

Torrance, John, ed. *The Concept of Nature*, The Herbert Spencer Lectures (Oxford: Clarendon Press, 1992).

Urmson, J. O. "On Grading", *Mind*, 59:234 (April 1950).

Vilkka, Leena. "Miltä tuntuu olla tikankontti," *Näkökulma yhteiskuntatieteelliseen ympäristötutkimukseen*, eds. Terttu Pakarinen, Leena Vilkka, and Eija Moisseinen, Acta Universitatis Tamperensis, ser. B, 37 (Tampere: Tampereen yliopisto, 1991).

———. *Ympäristöetiikka: Vastuu luonnosta, eläimistä ja tulevista sukupolvista* (Helsinki: Helsinki University Press, 1993).

———. "A New Dimension of Animal Rights," *Satya* (May 1995).

———. "On Animal Consciousness," *New Directions in Cognitive Science*, eds. Paavo Pylkkänen and Pauli Pylkkö (Helsinki: Finnish Artificial Intelligence Society, 1995).

Warren, Mary Anne. "The Rights of the Nonhuman World," *Environmental Philosophy: A Collection of Readings*, eds. Robert Elliot and Arran Gare (Milton Keynes: The Open University Press, 1983).

Wemelsfelder, Françoise. "Animal Boredom: Do Animals Miss What They've Never Known," *Euroniche Conference Proceedings: Animal Use in Education*, eds. Bryony Close, Francine Dolins, and Georgia Mason (London: Humane Education Centre, 1989).

———. *Animal Boredom: Towards an Empirical Approach of Animal Subjectivity* (The Netherlands: University of Leiden, Academic Dissertation, 1993).

Weston, Anthony. "Beyond Intrinsic Value: Pragmatism in Environmental Ethics," *Environmental Ethics*, 7:4 (1985).

Westra, Laura. *An Environmental Proposal for Ethics: The Principle of Integrity* (London: Rowman and Littlefield, 1994).

Wiepkema, P. R. and J. M. Koolhaas. "The Emotional Brain," *Animal Welfare*, 1 (1992).

Wilson, E. O. *The Diversity of Life* (London: Penguin Books, 1992).

———, ed. *Biodiversity* (Washington: National Academy Press, 1988).

von Wright, G. H. *The Logic of Preference: An Essay* (Edinburgh: Edinburgh University Press, 1963).

———. *The Varieties of Goodness* (London: Routledge and Kegan Paul, 1963).

———. *Humanismi elämänasenteena* (Helsinki: Otava, 1983).

———. "Arvot ja tarpeet," *Riittääkö energia, riittääkö järki*, eds. Pentti Malaska, Ismo Kantola, and Pirkko Kasanen (Helsinki: Gaudeamus, 1989).

Ziff, Paul. *Semantic Analysis* (New York: Cornell University Press, 1960).

Zimmerman, Michael. "On the Intrinsic Value of States of Pleasure," *Philosophy and Phenomenological Research*, 41 (September 1980-June 1981).

GLOSSARY

A

Anthropocentric interests
- Human interests centered on human values and welfare.

Anthropocentric value
- Opposed to naturocentric value.
- Value centered on people.
- Close to human value.

Anthropocentrism
- A philosophy according to which people are prior to the nonhuman world.
- Concerned with the topics of human values.

Anthropogenic beauty in nature
- People valuing some natural area as a beautiful place.

Anthropogenic value
- Value generated by people; value created or originated in the valuation processes of human beings.
- If we say that nature has anthropogenic value, this means value given or attributed to nature by people.

Axiological extensionism
- Involves both ethical and ontological extensionism of value.

Axiological goodness (in nature)
- A system of the forms of goodness in nature.

Axiology
- A theory or science of value.
- Subdivided into conceptual, ontological, epistemological, and normative research on value.

B

Biocentric value
- Value centered on a phenomenon of life.
- Example: biodiversity.

Biocentrism
- A philosophy centered on the topics of life.
- Example: Paul W. Taylor's biocentric ethics.

Biogenic beauty
- Beauty rooted in the phenomena of life.

Biogenic value
- Value produced by life; value created or originated in the processes of life.
- Example: survival value.

Biographical life
- People and animals have biographical life, if they have past and future, hopes and fears concerning their own lives in the historical sense.

Biological goodness
- Based on the biological interests of living organisms, plants, and animals.

Biological interests
- Based on the species-specific characteristics of an organism.

Biological life
- X has biological life, if X is alive.

C

Comparative statement of value
- Synonym: the topic of priorities.
- A basic form: X has more value K than Y.
- A question: Do animals have more intrinsic value than plants?

Connatural world
- "Mitwelt" (in German), implies that we and other species live together

within the world. (Klaus Meyer-Abich's term.)

Cultural constructivism
- A theory of value according to which all values are constructed by people within human culture.

D

Deep ecology
- Opposed to shallow ecology.
- The first principle of deep ecology is the well-being and flourishing of both human and nonhuman life on Earth.
- Originates in Arne Næss's environmental philosophy.

E

Ecocentric value
- Value centered on ecosystems.

Ecocentrism
- A philosophy centered on the topics of ecosystems and their well-being.
- Example: Aldo Leopold's land ethics.

Ecofeminism
- A synthesis of ecologically oriented and feminist philosophies.

Ecogenic beauty
- A part of ecosystem-based-aesthetics.

Ecogenic value
- Value produced by ecosystems; value created or originated in ecosystems.

Ecohumanism
- A philosophy aimed to relate ecological and humanistic principles of value.
- Based on a weak form of anthropocentrism.
- Example: Henryk Skolimowski's ecophilosophy.

Ecological imperative
- Act in such a way that you treat nature never solely as a means but always at the same time as an end.
- Kantian categorical imperative extended to nature.

Élan vital (Henri Bergson)
- Life involving intrinsic power to live.
- Close to will to live.

Ends-value
- Something has ends-value, if it is valued for its own sake.
- Opposed to means-value.
- Animals, living beings, and ecosystems have ends-value, if they are valued for their own sake.

Environmentalism
- The ideology of ecology as a movement stressing environmental problems and the ecological crisis.

Ethical extensionism
- Extends the scope of morality to include animals, living beings, and ecosystems, in addition to people.
- Classified to zoocentrism, biocentrism, and ecocentrism.

Existence-value
- Goodness in beingness.
- In environmental economics, the existence value of nature is understood in terms of ends-value, indicating people's willingness to preserve nature as such.

Experienced value
- sensed value, a felt or perceived quality of experience.

Extrinsic value
- X has extrinsic value, if X satisfies external or outside requirements for value.
- Examples: instrumental and transformative value.

G

Gaia hypothesis (J. E. Lovelock)
- All living beings are parts of Gaia which has the power to maintain and regulate the life-supporting systems of the Earth.

Genealogy of value
- Inquires into the origin and forms of value.

Good
- Synonym: good-in-itself.
- Opposed to "good for."
- Close to value.

Goodness
- A form of value.
- Subdivided into intrinsic, extrinsic, and systemic goodness.

H

Holism
- A philosophy centered on the topics of wholeness.
- A theory of value according to which the value of wholeness is not reducible to the value of individuals.
- opposed to individualism.
- opposed to universalism.
- A theory of value that places primal value on (bio)communities (like species, societies, ecosystems) instead of individuals or universals.
- Examples: J. Baird Callicott's anthropogenic holism, Klaus Michael Meyer-Abich's physiocentrism.

Human value
- Opposed to natural value.

I

Indexical intrinsic value
- Referred, mind-dependent value. (Robert Elliot's term of anthropogenic intrinsic value).

Individualism
- A philosophy centered on the topics of individuals and their well-being.
- A theory of value that places value on individuals as opposed to holism and universalism.

Inherent value$_L$
- Inherent value according to C. I. Lewis.
- A value attributed to objects that are intrinsically valued within a human brain.
- Example: the objects of arts, or a beautiful landscape may have inherent value$_L$.

Inherent value$_R$
- Inherent value according to Tom Regan.
- Close to zoogenic value.

Inherent worth$_T$
- Inherent worth according to Paul W. Taylor.
- Close to biogenic value.

Inner value
- Value in a thing, non-derivative value.
- Objects, things, or states have inner value, if they have value in themselves independent of other things.

Instrumental goodness
- An object has instrumental good-

ness, if it is a good tool or resource.
- A basic form: X as a tool or a resource is good for Y.
- Close to instrumental value.

Instrumental value
- An object has instrumental value, if it is used as a tool or resource.
- A basic form: X as a tool or resource has value for Y

Instrumental value view of nature
- Opposed to the intrinsic value view of nature.
- Involves the view of nature as to be utilized for several human purposes.
- Subdivided into wide and narrow instrumentalist views.

Interests
- Two distinctions: welfare and preference interests; anthropocentric and naturocentric interests.

Intrinsic goodness
- X has intrinsic goodness, if X is good in itself.

Intrinsic valuation
- A state with sympathy and love, the state of being fully involved in what you are doing or appreciating. (According to Robert Hartman).

Intrinsic value
- Opposed to intrinsic valuation.
- Opposed to extrinsic value.
- In terms of ends value opposed to means value.
- Classified in four basic forms: intrinsic value in terms of experienced value, ends-value, self-value, and inner value.
- A general definition: X has intrinsic value K on the grounds L from the viewpoint M.
- A zoocentric example: An animal has intrinsic value in terms of self-value on the grounds of having ability to value its own life from the viewpoint of its conscious, sentient nature.

Intrinsic value$_L$
- Intrinsic value according to C. I. Lewis.
- The state of pleasurable experiences in human brain.
- Synonym: experienced value.

Intrinsic-value view of nature
- Opposed to the instrumental-value view of nature.
- Involves the view of nature as needed to be respected and preserved as such.

Isolated value
- Value existing independent upon any other things.
- Opposed to related value.

L

Land ethics (Aldo Leopold)
- An ecocentrism according to which land should be regarded as a loved and respected community to which we belong rather than a commodity belonging to us.

Life
- Two forms: biological and biographical life.

M

Moral considerability
- A definitive topic: Can nature be morally considered?

Moral priority
- A comparative topic: Should animals have priority over plants?
- Example: An endangered plant

species has priority over a common species.

Moral significance
- A normative topic: Should X be morally considered to a particular degree?

Moral standing
- Questions: What is the scope of our morality? Which things should be taken into moral account?
- Subdivided into moral considerability, moral significance, and moral priority.

N

Narrow instrumentalist view of nature's value
- Nature has value solely in terms of money.

Natural beauty
- Subdivided into zoogenic, biogenic, and ecogenic beauty in nature.

Natural constructivism
- A theory of value according to which the crucial parts of value are constructed by nature within evolution, in addition to the values constructed by human culture.

Natural rights
- Synonym: intrinsic rights or naturogenic rights.
- Involves the assertion that people have rights on the grounds of their inner human nature.
- The extended assertion regards that animals should have rights on the grounds of their sentient nature and the other organisms on the grounds of their living nature.

Natural value
- Opposed to human value.

Nature
- Opposed to people.
- Opposed to culture.
- Grouped to animals, the phenomena of life, and ecosystems.

Nature conservation philosophy, synonym preservation
- Opposed to nature management.
- Attempts to form philosophical bases for nature preservation both on anthropocentric and naturocentric grounds.
- Inquires into human relations and attitudes to nature.

Naturistic theory of value
- A combination of animal, life, and ecosystem values.

Naturocentric interests
- Human interests centered on natural values and focused on the well-being of animals and nature.

Naturocentrism
- Opposed to anthropocentrism.
- Extended from anthropocentrism.
- Subdivided into zoocentrism, biocentrism, ecocentrism, and physiocentrism.

Naturogenic value
- Value generated by nature; value created or originated in nature.
- Opposed to anthropogenic value.
- Subdivided into zoogenic, biogenic, and ecogenic values.

"Naturopology"
- Opposed to biology.
- Compared with anthropology.
- Inquires into animals and plants on historical, cultural, and philosophical perspectives.

O

Objective value
- Value inhering in an object.

- A basic form of objective intrinsic value: X is value in itself.
- Subdivided into the weak and strong objective value views.

Ontological extensionism (of value)
- Extends the scope of value from people to animals, living beings, and ecosystems.
- Classified to zoogenism, biogenism, and ecogenism.

P

Physiocentric value
- Value centered on nature as a whole.

Physiocentrism
- A philosophy centered on the topics of nature as a whole.
- Example: Klaus Michael Meyer-Abich's philosophy.

Physiogenic value
- Value generated by nature as a whole; value created or originated in nature as a whole.

Pluralistic theory of value
- A theory according to which objective, relative, and subjective values are combinable together instead of perceiving them as conflicting values.

Preference interests
- Refers to that which an individual wants.

Q

Qualitative statement of value
- A basic form: X has value K.
- A question: Do animals have intrinsic value?

Quantitative statement of value
- A basic form: X's degree of value K is r.
- A question: How much have animals intrinsic value?

R

Relative value, related value
- Requires both a valued object and the valuing subject: Y values X. (Y and X emphasized equally.)

Rights of nature
- Based on the intrinsic value view of nature.
- Rights as anthropocentric are centered on people, correspondingly rights as naturocentric are centered on animals and nature.
- Moral and legal rights as anthropogenic can be given to nonhuman entities.
- See natural rights.

S

Self value
- Involves the Kantian idea that each human person is indispensable as such. Extended to animals who are the subjects of their own lives.

Strong anthropocentrism
- According to this view, no intrinsic value is allowed to be attributed to the non-human world.

Strong objective-value view
- Argues for mind-independent values. Value is ontologically independent upon human valuation.

Subjective value
- Value that exclusively depends on a subject, to whom it owns its existence.
- A basic form of subjective intrinsic value: Y values X for its own sake.

Subjectivistic fallacy

- A confusion between values and valuers.

Systemic value
- Synonym: systemic goodness.
- X has systemic value, if X satisfies functional or processive requirements for value.

T

Theory of animal value
- A combination of the zoocentric and zoogenic topics of value.

Theory of ecosystem value
- A combination of the ecocentric and ecogenic topics of value.

Theory of life value
- A combination of the biocentric and biogenic topics of value.

Theory of the intrinsic value of nature
- Based on axiological extensionism and value pluralism.
- Develops a naturocentric and naturogenic view of value.

Transformative value
- Synonym: contributive value.
- An object has transformative value, if it has value for the sake of the valuable things derived from it. Value is given to the object, because it has valuable consequences to other objects or it otherwise contributes to the realization of value.
- A basic form: X as such has value for Y.

Truncated intrinsic value
- Attributed, mind-dependent, anthropogenic value. (J. Baird Callicott's term).

U

Universalism
- A philosophy of value regarding values as universals.
- Opposed to individualism.
- Classic universal values: goodness, beauty, truth.

V

Value, synonym worth
- Opposed to valuation.
- Close to good.

Value-objectivism
- A theory giving priority to values over valuation. Values determine valuation processes: No value, no valuations.

Value-relativism
- An interactive theory of value. Both are needed in creation of intrinsic value: objective values and subjective valuing.'Value-subjectivism
- A theory of value emphasizing the meaning of valuation: No valuation, no values.

W

Weak anthropocentrism
- According to this view, people are allowed to attribute intrinsic value in terms of ends value to the nonhuman world, albeit people have superior value to other beings.

Weak objective-value view
- Valuation is impersonal and intersubjective: my personal opinion does not determine my valuation. Values should be reasoned by well-grounded arguments and general axiological rules.

Welfare interest
- Refers to an organism's state of well-being.
- Close to biological interests.

Wide instrumentalist view of nature's value
- Accepts that nature has several types of instrumental value: economic, aesthetic, scientific, recreational, religious value, and so on.

Will to live (Arthur Schopenhauer)
- The force of nature; that which acts and strives in nature.
- Close to *élan vital.*

Z

Zoocentric value
- Value centered on animals.
- Example: sympathy with animals.

Zoocentrism
- A philosophy centered on the topics of animals and their welfare.
- Example: Animal rights philosophy.

Zoogenic beauty
- Beauty rooted in animals, animal aesthetics.

Zoogenic value
- Value generated by animals; value created or originated in the valuation processes of animals.
- Example: the well-being of animals.

ABOUT THE AUTHOR

Leena Vilkka, Ph.D., is Researcher at the Academy of Finland and University Lecturer in Environmental Philosophy. She is the author, among other works, of *Ympäristöetiikka* (Environmental Ethics) and *Eläinten Tietoisuus ja Oikeudet* (Animal Consciousness and Rights). Vilkka is also an environmental activist. She chairs two Finnish organizations, the Wolf Group and Green Union for the Protection of Life. Her address is Department of Philosophy, 00014 University of Helsinki, Finland.

INDEX

VIBS

The **Value Inquiry Book Series** is co-sponsored by:

American Maritain Association
American Society for Value Inquiry
Association for Personalist Studies
Association for Process Philosophy of Education
Center for East European Dialogue and Development, Rochester Institute of Technology
Centre for Cultural Research, Aarhus University
College of Education and Allied Professions, Bowling Green State University
Concerned Philosophers for Peace
Conference of Philosophical Societies
International Academy of Philosophy of the Principality of Liechtenstein
International Society for Universalism
International Society for Value Inquiry
Natural Law Society
Philosophical Society of Finland
Philosophy Born of Struggle Association
Philosophy Seminar, University of Mainz
R.S. Hartman Institute for Formal and Applied Axiology
Society for Iberian and Latin-American Thought
Society for the Philosophic Study of Genocide and the Holocaust
Society for the Philosophy of Sex and Love
Yves R. Simon Institute.

Titles Published

1. Noel Balzer, *The Human Being as a Logical Thinker.*

2. Archie J. Bahm, *Axiology: The Science of Values.*

3. H. P. P. (Hennie) Lötter, *Justice for an Unjust Society.*

4. H. G. Callaway, *Context for Meaning and Analysis: A Critical Study in the Philosophy of Language.*

5. Benjamin S. Llamzon, *A Humane Case for Moral Intuition.*

6. James R. Watson, *Between Auschwitz and Tradition: Postmodern Reflections on the Task of Thinking.* A volume in **Holocaust and Genocide Studies.**

7. Robert S. Hartman, *Freedom to Live: The Robert Hartman Story,* edited by Arthur R. Ellis. A volume in **Hartman Institute Axiology Studies.**

8. Archie J. Bahm, *Ethics: The Science of Oughtness.*

9. George David Miller, *An Idiosyncratic Ethics; Or, the Lauramachean Ethics.*

10. Joseph P. DeMarco, *A Coherence Theory in Ethics.*

11. Frank G. Forrest, *Valuemetrics: The Science of Personal and Professional Ethics.* A volume in **Hartman Institute Axiology Studies.**

12. William Gerber, *The Meaning of Life: Insights of the World's Great Thinkers.*

13. Richard T. Hull, Editor, *A Quarter Century of Value Inquiry: Presidential Addresses of the American Society for Value Inquiry.* A volume in **Histories and Addresses of Philosophical Societies.**

14. William Gerber, *Nuggets of Wisdom from Great Jewish Thinkers: From Biblical Times to the Present.*

15. Sidney Axinn, *The Logic of Hope: Extensions of Kant's View of Religion.*

16. Messay Kebede, *Meaning and Development.*

17. Amihud Gilead, *The Platonic Odyssey: A Philosophical-Literary Inquiry into the* Phaedo.

18. Necip Fikri Alican, *Mill's Principle of Utility: A Defense of John Stuart Mill's Notorious Proof.* A volume in **Universal Justice.**

19. Michael H. Mitias, Editor, *Philosophy and Architecture.*

20. Roger T. Simonds, *Rational Individualism: The Perennial Philosophy of Legal Interpretation.* A volume in **Natural Law Studies.**

21. William Pencak, *The Conflict of Law and Justice in the Icelandic Sagas.*

22. Samuel M. Natale and Brian M. Rothschild, Editors, *Values, Work, Education: The Meanings of Work.*

23. N. Georgopoulos and Michael Heim, Editors, *Being Human in the Ultimate: Studies in the Thought of John M. Anderson.*

24. Robert Wesson and Patricia A. Williams, Editors, *Evolution and Human Values.*

25. Wim J. van der Steen, *Facts, Values, and Methodology: A New Approach to Ethics.*

26. Avi Sagi and Daniel Statman, *Religion and Morality.*

27. Albert William Levi, *The High Road of Humanity: The Seven Ethical Ages of Western Man,* edited by Donald Phillip Verene and Molly Black Verene.

28. Samuel M. Natale and Brian M. Rothschild, Editors, *Work Values: Education, Organization, and Religious Concerns.*

29. Laurence F. Bove and Laura Duhan Kaplan, Editors, *From the Eye of the Storm: Regional Conflicts and the Philosophy of Peace.* A volume in **Philosophy of Peace.**

30. Robin Attfield, *Value, Obligation, and Meta-Ethics.*

31. William Gerber, *The Deepest Questions You Can Ask About God: As Answered by the World's Great Thinkers.*

32. Daniel Statman, *Moral Dilemmas.*

33. Rem B. Edwards, Editor, *Formal Axiology and Its Critics.* A volume in **Hartman Institute Axiology Studies.**

34. George David Miller and Conrad P. Pritscher, *On Education and Values: In Praise of Pariahs and Nomads.* A volume in **Philosophy of Education.**

35. Paul S. Penner, *Altruistic Behavior: An Inquiry into Motivation.*

36. Corbin Fowler, *Morality for Moderns.*

37. Giambattista Vico, *The Art of Rhetoric* (*Institutiones Oratoriae,* 1711-1741), from the definitive Latin text and notes, Italian commentary and introduction by Giuliano Crifò, translated and edited by Giorgio A. Pinton and Arthur W. Shippee. A volume in **Values in Italian Philosophy.**

38. W. H. Werkmeister, *Martin Heidegger on the Way,* edited by Richard T. Hull. A volume in **Werkmeister Studies.**

39. Phillip Stambovsky, *Myth and the Limits of Reason.*

40. Samantha Brennan, Tracy Isaacs, and Michael Milde, Editors, *A Question of Values: New Canadian Perspectives in Ethics and Political Philosophy.*

41. Peter A. Redpath, *Cartesian Nightmare: An Introduction to Transcendental Sophistry.* A volume in **Studies in the History of Western Philosophy.**

42. Clark Butler, *History as the Story of Freedom: Philosophy in Intercultural Context,* with Responses by sixteen scholars.

43. Dennis Rohatyn, *Philosophy History Sophistry.*

44. Leon Shaskolsky Sheleff, *Social Cohesion and Legal Coercion: A Critique of Weber, Durkheim, and Marx.* Afterword by Virginia Black.

45. Alan Soble, Editor, *Sex, Love, and Friendship: Studies of the Society for the Philosophy of Sex and Love, 1977-1992.* A volume in **Histories and Addresses of Philosophical Societies.**

46. Peter A. Redpath, *Wisdom's Odyssey: From Philosophy to Transcendental Sophistry.* A volume in **Studies in the History of Western Philosophy.**

47. Albert A. Anderson, *Universal Justice: A Dialectical Approach.* A volume in **Universal Justice.**

48. Pio Colonnello, *The Philosophy of José Gaos.* Translated from Italian by Peter Cocozzella. Edited by Myra Moss. Introduction by Giovanni Gullace. A volume in **Values in Italian Philosophy.**

49. Laura Duhan Kaplan and Laurence F. Bove, Editors, *Philosophical Perspectives on Power and Domination: Theories and Practices.* A volume in **Philosophy of Peace.**

50. Gregory F. Mellema, *Collective Responsibility.*

51. Josef Seifert, *What Is Life? The Originality, Irreducibility, and Value of Life.* A volume in **Central-European Value Studies.**

52. William Gerber, *Anatomy of What We Value Most.*

53. Armando Molina, *Our Ways: Values and Character,* edited by Rem B. Edwards. A volume in **Hartman Institute Axiology Studies.**

54. Kathleen J. Wininger, *Nietzsche's Reclamation of Philosophy.* A volume in **Central-European Value Studies.**

55. Thomas Magnell, Editor, *Explorations of Value.*

56. HPP (Hennie) Lötter, *Injustice, Violence, and Peace: The Case of South Africa.* A volume in **Philosophy of Peace.**

57. Lennart Nordenfelt, *Talking About Health: A Philosophical Dialogue.* A volume in **Nordic Value Studies.**

58. Jon Mills and Janusz A. Polanowski, *The Ontology of Prejudice.* A volume in **Philosophy and Psychology.**

59. Leena Vilkka, *The Intrinsic Value of Nature.*